The Politics of Mental Health and Illness

Series Editor
Bruce M. Z. Cohen
Sociology
University of Auckland
Auckland, New Zealand

The Politics of Mental Health and Illness conceptualises the western mental health system as a social, economic, political, and cultural project which cannot be adequately theorised without considering wider societal and structural issues such as professional power, labelling and deviance, ideological and social control, consumption, capital, and self-governance. Engaging with both social theory and empirical evidence, and situated at the intersections of critical psychiatry, sociology, anthropology, social psychology, social work, history, media studies, and the health and education sciences, the series offers academics, researchers, postgraduate students, and practitioners new and innovative monographs and edited collections with which to contest and challenge the taken-for-granted understandings of psychiatry and the mental health system currently progressed in western society. Research topics for the collection may include: aspects of the medicalisation and/or pharmaceuticalisation of everyday life; psy-professional power and knowledge production; psychopolitics and the expansion of talk therapies; the biomedical model and the future of psychiatric nosology; the marketisation of mental health discourse; social media and self-diagnosing behaviour; capitalism and psychiatric violence; and specific analyses of psy-professionals and their practices from a variety of locations (including prisons, workplaces, addiction clinics, the military, schools, the home, and general hospitals). As the first book series to explicitly encourage critical contributions in the area of mental health and illness, we welcome scholarship from a wide range of different theoretical and empirical perspectives.

If you would like to discuss a book idea prior to submitting a proposal please contact the Series Editor, Dr Bruce M. Z. Cohen, via the Commissioning Editor, Beth Farrow (beth.farrow@palgrave.com).

Natalie Ann Hendry • Effie Karageorgos
Editors

Critical Mental Health in Australia and Aotearoa New Zealand

Social and Historical Perspectives

palgrave
macmillan

Editors
Natalie Ann Hendry
Faculty of Education
University of Melbourne
Melbourne, VIC, Australia

Effie Karageorgos
School of Humanities, Creative Industries
and Social Sciences
University of Newcastle
Newcastle, NSW, Australia

ISSN 2731-5266 ISSN 2731-5274 (electronic)
The Politics of Mental Health and Illness
ISBN 978-3-031-82860-7 ISBN 978-3-031-82861-4 (eBook)
https://doi.org/10.1007/978-3-031-82861-4

Cover illustration: Malte Mueller

This Palgrave Macmillan imprint is published by the registered company Springer Nature Switzerland AG.
The registered company address is: Gewerbestrasse 11, 6330 Cham, Switzerland

For dad—N.A.H.
For Janet—E.K.

Acknowledgments

We first thank the Indigenous groups on whose lands the events described in the chapters of this book, as well as the work carried out in researching, writing and organising, took place. In particular, we thank the Wurundjeri people of the Kulin nation, the Pambalong clan of the Awabakal people and the Gadigal people of the Eora nation, where the editors of this book live and work. We extend support and solidarity to all colonised peoples across the world today.

There is a large group of people without whom this book would not exist. We thank every person who has presented, attended or has otherwise been involved in the Social Production of Mental Health seminar series. We especially thank Baptiste Brossard for developing the series during the dark days of COVID lockdowns and leading it ably from 2020 to 2021. Very special thanks also go to Bruce M.Z. Cohen, who has supported the series since its early days and who is a supportive presence to all who work in and around critical mental health. Thank you for approaching us about publishing this collection of excellent chapters in The Politics of Mental Health and Illness series. We are also very grateful to be working with all of the authors, and the people and departments that support and enable their work. Doing 'critical mental health' work can be isolating at times, so thank you for collaborating with us.

We also have separate people and groups to thank.

Natalie Ann Hendry

The ever-constraining conditions of higher education in Australia continue to challenge and diminish spaces for collective conversation, reflection, research and writing. So, any opportunity to create a little pocket of sharing and thinking otherwise together is so very crucial. Thanks to Effie Karageorgos for sharing and sustaining this little pocket with me, in the seminar series, this book, on screens, and over delicious food in person.

Thanks to my colleagues for your support—all the incredible people, in too many different disciplines, departments, universities and organisations here and beyond Australia, that support doing critical work in so many different ways. Thanks especially to Kath Albury and Daniel Reeders for always unpacking those frustrating moments during research with me.

I also thank the young people, students, teachers, lecturers, peer workers, allied health staff, practitioners, clinicians, families and carers who offered me a sort of 'apprenticeship' in Australian mental health from my childhood onwards and encouraged me to ask *those* questions about my own ambivalent experiences and history. Thanks Sam for sustaining space and my energy for this work.

Effie Karageorgos

Thank you to everyone who welcomed the inclusion of a historian into the Social Production of Mental Health seminar series in 2022, and particularly Natalie Ann Hendry, who has been an excellent collaborator and friend. The series has been a constant source of inspiration and joy, and I'm very privileged to be able to share ideas with its speakers and participants.

Thanks to the State Library of New South Wales, Australian War Memorial and National Archives of Australia for always excellent help while accessing archives. Thank you to Marrickville Library for the writing space.

I have people to thank without whose help I wouldn't be able to write my chapter or edit this book. Catharine Coleborne and Hans Pols have been kind and generous mentors. I am lucky also to be surrounded weekly by a group of excellent historians and sociologists, with special thanks to Nancy Cushing, Kathleen McPhillips, Elizabeth Roberts-Pedersen and Robyn Dunlop. I wrote this chapter between Awabakal and Gadigal lands, including the University of Newcastle, State Library of

New South Wales, Marrickville Library and Lake Macquarie, and during this time benefited from conversations with and inspiration from some of the best people, including Christian Thompson, Zora Simic, Peter Polites, Dimitra Giannacopoulos and Karen Bird.

My interest in the study of war, violence and psychiatry has been greatly influenced by the experiences of the Kotrotsios, Vlachos, Dallas and Karageorgos families in the Greek Civil War and during the Greek military junta, as well as the both traumatic and not-so-traumatic migration experiences of some as settlers to Gadigal and Kaurna lands. Love to everybody, in particular Chris Karageorgos, Vicki Karageorgos, Lydia Karageorgos, Matt Pinson, Atticus Pinson, Jenny Farmacoulas, Voula Farmacoulas and Eleftheria Kotrotsios (Vlachos). A big hug to Peponi.

Contents

Notes on Contributors

Samantha Baker received her Honours degree from the School of History and Philosophy of Science at the University of Sydney in 2023 and is a member of the Re;Minding Histories research group. Her thesis was entitled 'A House of One's Own: The Politics of Identity and Australia's First Trans Refuge', for which she was awarded the John Dennis Kinkhead Prize in the History and Philosophy of Science. She is currently exploring opportunities for postgraduate studies.

Roslyn Burge is a Sydney-based public historian and a member of the Re;Minding Histories research group. Her work includes commissioned histories and exhibitions. She has a special interest in the history of mental health and that of Callan Park Mental Hospital. Oral history is significant in her practice, and it was much enriched by the co-production research of the Re;Minding Histories tribe.

Bruce M. Z. Cohen is an associate professor and a sociologist of mental health at the Waipapa Taumata Rau University of Auckland, and author of *Mental Health User Narratives: New Perspectives on Illness and Recovery* (Palgrave, 2008) and *Psychiatric Hegemony: A Marxist Theory of Mental Illness* (Palgrave, 2016). He is the Editor of the *Routledge International Handbook of Critical Mental Health* (Routledge, 2018) and, with Martin Harbusch and Jo Reichertz, the forthcoming Palgrave collection *Addiction* *and the Medicalisation of Conspicuous Behaviour: New Critical Perspectives.* Bruce is also the Editor of the Palgrave book series *The Politics of Mental Health and Illness.*

Catharine Coleborne is a professor at the School of Humanities, Creative Industries and Social Sciences at the University of Newcastle, and a member of the Re;Minding Histories research group. Her research explores the social and cultural histories of mental illness and the institutions created to confine, treat and assist the mentally ill.

Jacinthe Flore is Lecturer in History and Philosophy of Science in the School of Historical and Philosophical Studies at the University of Melbourne. She is the author of *A Genealogy of Appetite in the Sexual Sciences* (Palgrave Macmillan, 2020) and *The Artefacts of Digital Mental Health* (Palgrave Macmillan, 2023).

Anthony Harris is an academic psychiatrist at the University of Sydney and a member of the Re;Minding Histories research group. In his clinical role, he cares for young people with severe mental illness. His academic life includes psychophysiological and treatment research. He teaches medical students and health professionals and is active in the NGO sector.

Natalie Ann Hendry is an educator and researcher who has worked and collaborated across community, secondary and hospital-based education sectors since 2003. Forever curious about the messy relationships between health, digital media and education, Natalie's research engages with media studies, cultural studies and sociological theories, and uses digital ethnography, creative methods and policy analysis approaches. She currently is Senior Lecturer in Youth Wellbeing in the Faculty of Education at the University of Melbourne, where she teaches social research methodology and social and critical approaches to wellbeing in schools.

Rebecca Howe completed her PhD research at the University of Sydney in the School of Education and Social Work. With an extensive history working in the LGBTIQASB+ youth homelessness and health policy sectors, Rebecca currently works at the University of Melbourne, where she coordinates the Australian Centre Graduate Research Program in Indigenous Settler Relations.

Effie Karageorgos is Senior Lecturer in History at the University of Newcastle (UON). Her research focuses on the social history of war through the study of protest, violence, gender and psychiatry. She is the Deputy Co-Director of the UON Centre for Society, Health and Care Research, and Editor of *Health and History*, the official journal of the Australian and New Zealand Society for the History of Medicine. She runs the Social Production of Mental Health seminar series alongside Dr Natalie Ann Hendry.

Holly Kemp is employed as a peer support worker at NSW Health and a member of the Re;Minding Histories research group.

Diana Kopua Te Whare Wānanga o Te Kurahuna; MBChB; Fellow RANZCP, Iwi: Ngāti Porou, is a Fellow of the Royal Australia New Zealand College of Psychiatry. As the first Ngāti Porou psychiatrist, she forged space for the work she had started as a nurse—Mahi a Atua, an indigenous approach developed in the mid-1990s to address ongoing issues of institutional racism and inequity for Māori within New Zealand's health services and society.

Mark Kopua Te Whare Wānanga o Te Kurahuna; Tohunga, Iwi: Te Aitanga a Hauiti, Ngāti Ira, Ngāti Porou, is a historian, master carver, trailblazer in modern Moko (traditional Māori tattoo) and design consultant. His approach to healing, rooted in storytelling and ancient Māori knowledge, has resonated deeply with communities. In partnership with his wife, Dr. Diana Kopua, Mark co-founded 'Te Kurahuna,' a whare wānanga (training institute) dedicated to fostering indigenous knowledge through unique and authentic learning experiences. Together, they have created a space where practitioners can engage with traditional Māori wisdom, ensuring its preservation and continuity for future generations.

Michelle Levy Te Whare Wānanga o Te Kurahuna; Mataora/Kaupapa Māori researcher/writer, Iwi: Waikato, Ngāti Mahuta, has a PhD in community psychology from the University of Waikato. When she is not working with Te Kurahuna to illuminate the critical theory and evidence base of Mahi a Atua, you can find her educating, creating and activating as part of the Toitū te Tiriti Whāingaroa community-led kaupapa; playing in her art studio; or walking her rescue dogs on the black sand beach of Whāingaroa.

Roberto McLeay is a Professional Teaching Fellow and Doctoral Candidate in the Faculty of Education and Social Work, University of Auckland, Aotearoa New Zealand. Roberto's background in primary school teaching, pastoral care and counselling has led him to focus on researching children's emotions in Aotearoa New Zealand schools, with a unique interest in how current emotions, mental health and wellbeing are used to shape young people's understandings of themselves and others.

Hans Pols is a professor at the School of History and Philosophy of Science at the University of Sydney, and a member of the Re;Minding Histories research group. He has a long-standing interest in the history of psychiatry and mental health care. Over the past ten years, he has increasingly become involved in addressing the recent history and possible futures of mental health services in Indonesia and Australia.

Paul Rhodes is Professor of Clinical Psychology at the University of Sydney and a member of the Re;Minding Histories research group. He is also co-convenor of the Ecological Emotions Research Lab and practicing artist at Lennox St Studios.

Aimee Sinclair is a PhD candidate at Curtin University. Her work considers the effects of including peer support work as an occupation within the Australian mental system, drawing on Mad and relational ways of thinking-feeling-doing.

Gemma Lucy Smart is a lived experience PhD scholar at the University of Sydney. She is currently finishing her dissertation on spirituality and critical perspectives in psychiatry during late twentieth-century Australia with the Re;Minding Histories research group. She's known for her philosophy of psychiatry research on Internet Gaming Disorder. She's a disability activist, musician, and poet.

Kris Taylor has a PhD in Psychology from the University of Auckland. His work turns a critical eye towards pornography viewership and the concept of pornography addiction, as well as the field of pornography research in psychology more generally. His current project works with secondary school age boys and young men, encouraging recognition and critique of peer group tolerance for sexual harassment and abuse.

Emma Tseris works on unceded Gadigal Land and is Senior Lecturer in Social Work and Policy Studies at the University of Sydney, Australia. Her research is focused on critical mental health theory, and critical feminist analyses of mental health discourses and systems. Emma is engaged in multiple projects investigating the intersections between gender inequality and biomedical understandings of human experience.

Georgia Valis received her Honours degree from the School of History and Philosophy of Science at the University of Sydney in 2022 and is a member of the Re;Minding Histories research group. Her thesis is entitled 'Isolated and Invisible: Migration and Psychiatry in Post-War Australia'. She is currently exploring future career options.

Eloise Watts received her Honours degree from the School of History and Philosophy of Science at the University of Sydney in 2023 and is a member of the Re;Minding Histories research group. Her thesis was entitled 'Louisa Lawson House: A Feminist Mental Health Service in 1980s Sydney'. Upon graduation, she received the university medal. She is currently exploring which career options will allow her to realise her ideals best.

Asha Zappa is currently employed as a peer worker by Caritas Acute Mental Health Unit and a member of the Re;Minding Histories research group. They are a passionate and committed art therapist.

Abbreviations

ADF	Australian Defence Force
ADHD	attention-deficit/hyperactivity disorder
AIHW	Australian Institute of Health and Welfare
AOA	American Orthopsychiatric Association
APA	American Psychiatric Association
APA	American Psychiatric Association
APS	Australian Psychological Society
ARC	Australian Research Council
C/S/X	Consumer/Survivor/Ex-Patient
C/S/X	consumer/survivor/ex-patient
C/S/X	consumer/survivor/ex-patient
CASEL	Collaborative for Academic, Social, and Emotional Learning
CI	Chief Investigator
DMHWS	Defence Mental Health and Wellbeing Strategy
DSM	Diagnostic and Statistical Manual of Mental Disorders
DSM	Diagnostic and Statistical Manual of Mental Disorders
DSM	Diagnostic and Statistical Manual of Mental Disorders
DSM	Diagnostic and Statistical Manual of Mental Disorders
DVA	Department of Veterans' Affairs
EC	emotional competence
ECT	electroconvulsive therapy
EI	emotional intelligence
FOMN	Future of Madness Network

GATE	Global Action for Trans Equality
HBIGDA	Harry Benjamin International Gender Dysphoria Association
ICD	International Classification of Diseases
ICD	International Classification of Diseases
LGBT	Lesbian Gay Bisexual Transgender
LGBTI	Lesbian Gay Bisexual Transgender Intersex
LGBTQIA+	Lesbian, Gay, Bisexual, Transgender, Queer, Intersex, Asexual
MILF	mother I'd like to fuck
NCC	National Council of Churches
NGO	Non-Government Organisation
OECD	Organisation for Economic Co-operation and Development
PI	Partner Investigator
PTSD	post-traumatic stress disorder
RA	Research Assistant
RAAF	Royal Australian Air Force
RMO	Regimental Medical Officer
SEL	social and emotional learning
SOC	Standards of Care
SOGIESC	sexual orientation, gender identity, gender expression, sex characteristics
SPMH	Social Production of Mental Health
SPMH	Social Production of Mental Health
STEM	Science, Technology, Engineering and Mathematics
UK	United Kingdom
UN	United Nations
VA	Veteran's Administration
VVAW	Vietnam Veterans Against the War
VVWG	Vietnam Veterans Working Group
WPATH	World Professional Association of Transgender Health
YP	Yogyakarta Principles

1

Introduction: Remaking Experiences and Knowledges Through Critical Mental Health Research in Australia and Aotearoa New Zealand

Natalie Ann Hendry and Effie Karageorgos

Introduction

As the concepts of 'mental health' and 'mental illness' have broadened, multi-disciplinary researchers have re-examined the place of psychology and psychiatry in modern society. Their analyses typically apply social and historical ideas about how mental health/illness is constituted to examine the pervasive influence of psy-considerations and concepts in everyday lives. Despite changes in the approaches to mental health, the tendency towards a biological (or now, increasingly a neuro-biological),

N. A. Hendry (✉)
Faculty of Education, University of Melbourne, Melbourne, VIC, Australia
e-mail: natalie.hendry@unimelb.edu.au

E. Karageorgos
School of Humanities, Creative Industries and Social Sciences, University of Newcastle, Newcastle, NSW, Australia
e-mail: effie.karageorgos@newcastle.edu.au

N. A. Hendry, E. Karageorgos (eds.), *Critical Mental Health in Australia and Aotearoa New Zealand*, The Politics of Mental Health and Illness,
https://doi.org/10.1007/978-3-031-82861-4_1

Kraepelian view of mental illness has persisted into the twenty-first century, particularly demonstrated by the publication of DSM-5 (American Psychiatric Association, 2013). The resulting expansion of psychiatric symptoms and disorders within this fifth volume has variously contributed to the pathologisation of everyday life, which has prompted a series of responses by scholars from a range of disciplines. We can also see how this mundane expansion of the psy-disciplines into supposedly more humane and transformational domains—digital technology, schooling, human resources and social media, for example—has not transformed the relationships between communities, practices, professionals and institutions that often replicate and reinforce long-standing histories of discrimination and violence motivated by such psychiatric classification and treatment.

The diversity of interpretations relating to the presence, absence or critique of mental health and illness in human societies signal the seeming universality of such issues. However, within the academe there is frequently little interaction between disciplines on these topics, despite similarities in approaches, methods, insights, applications and, importantly, politics. In response to this relative scarcity, *Critical Mental Health in Australia and Aotearoa New Zealand: Social and Historical Perspectives* represents just some of the vital, current diversity within academic studies of mental health and illness in Australia and Aotearoa New Zealand.

Many of the chapters in this volume have emerged from the Social Production of Mental Health (SPMH) interdisciplinary seminar series which, since 2020, has featured scholars working in fields such as sociology, history, education, social work, physiotherapy and psychology, reflecting the increased criticality within social and historical perspectives of psychiatry, psychology and mental health. The SPMH seminar series was devised and facilitated by sociologist Baptiste Brossard (formerly Australian National University) in the midst of the COVID-19 lockdowns in 2020. As editors of this collection, we took up facilitation of the SPMH series as Brossard departed for the University of York at the end of 2021: media and education scholar Natalie Ann Hendry (formerly RMIT University, now The University of Melbourne) joined Brossard in facilitating the series at the start of 2021, and historian Effie Karageorgos (The University of Newcastle) came on board to co-convene the series with Hendry and extend its interdisciplinarity from 2022 onwards. Our book series editor, sociologist Bruce M. Z. Cohen (The University of Auckland), was an enthusiastic participant in the inaugural and subsequent seminar series, and conversations about this book evolved over the last few years' seminars. Karageorgos' entry also saw the series formally associated with the interdisciplinary Future of Madness Network (FOMN) at the University of Newcastle, which was created by Professor Catharine Coleborne, co-convened with Karageorgos, and since 2025 expanded to form the UON Centre for Society, Health and Care Research, co-directed by Professor Coleborne and Associate Professor Caragh Brosnan, with Karageorgos and Associate Professor Penny Buykx serving as Deputy Co-Directors.

The presentations and conversations shared over the life of the series have spoken to the similarities—and important differences—across diverse fields when relating these perspectives to specific disciplinary approaches and problems. They have also pointed to the necessary interdisciplinarity within broader studies of psychiatry and mental health. *Critical Mental Health* addresses the existing—often unrecognised—overlap between disciplines that traditionally draw on similar arguments and point to shared relationships within mental health 'care'. What stands out to both of us as editors and co-convenors is that on the surface it can seem as if not much is happening as a *collective* critical mental health scholarship in both Australia and Aotearoa New Zealand. Yet there are substantial studies and relationships building across disciplines, institutions and methodologies, as well as within and outside of the academe across this region. This collection aims to offer an insight into such research.

Like other books, collections and series oriented towards the critical, it almost seems to be expected that we should carefully articulate our approach towards criticality: to define what critical means for this collection, and what theoretical and empirical work criticality 'does' or what we hope it will do. As Cohen (2018a) outlines in his own introduction to 'critical' in the *Routledge International Handbook of Critical Mental Health*, while critical mental health research is diverse and heterogeneous, this fragmented field has a shared history emerging from anti-psychiatric, social constructivist, labelling theory, structural functionalist and Foucauldian approaches. But what is perhaps more important here for us, is that like Cohen (2018a), we acknowledge how research and theory within higher education continues to be steered towards supposedly pragmatic rather than critical scholarship. We can see this in research funding opportunities that narrow the scope of theoretical inquiry or teaching related to mental health (see Chap. 3, Tseris) or claims that criticality offers no practical guidance for change and is consequently futile. And yet, the SPMH seminar series, FOMN Network and our contributors' other public initiatives and activisms have offered crucial space and time for thinking otherwise for practitioners, peer workers and others 'in' the field. Historical accounts, too, have shown us how much remains unchanged within the mental health industry, therapeutic cultures, psychological fields and psychiatric domains in both Australia and Aotearoa New Zealand. Given this stasis and wariness towards the critical, this collection centres approaches that do not presume that psychology and psychiatry are merely benign nor inherently benevolent undertakings. We understand that good intentions are insufficient—a point that we are gratefully reminded of again and again from colleagues, peers, allies and leaders in Mad Studies and psy-survivor movements.

Historicising Critical Mental Health

As our chapters demonstrate, it is difficult to cleave apart historical perspectives from social perspectives. In writing a historiography or a review of historical and sociological perspectives on mental health and psychiatry, the truly interdisciplinary nature of this field becomes evident. Historians have adopted sociological theories or techniques to analyse the past, and sociologists examine the historical context of current issues related to mental health. Our experience in convening the SPMH seminar series has demonstrated clearly to us that we need to look far beyond these disciplines to understand the breadth of scholarly attention on these themes—what is universal here is not mental health experiences themselves but rather that mental health is a universally important issue in world societies.

Critical mental health studies, not only in Australia and Aotearoa New Zealand but across many parts of the world, have emerged from both an interdisciplinary tradition and a focus on studies that necessarily centre gender, class, race and colonialism—factors that play a key role in decisions made about and experiences of mental health care (see Chap. 7, Kopua et al., and Chap. 10, Howe, for example). Histories of psychiatry and mental health have necessarily addressed these themes to differing degrees—placing experiences of care within social and cultural contexts of the past. Largely, such studies emerging from Australia and Aotearoa New Zealand have moved forward in similar ways to research conducted in other parts of the world but have been influenced by their specific circumstances from the colonial period to the current day.

Although sociologists from Australia and Aotearoa New Zealand were not immune to the influence of anti-psychiatry in recasting the history of institutionalisation as one primarily concerned with deviance, power and control in the 1960s and 1970s (Cohen, 2018a; Foucault, 1964; Laing, 1965; Scheff, 1963; Szasz, 1974), work carried out by historians did not incorporate such ideas until the following decades. Key to the 'social turn' in general histories of psychiatry are scholars Mark Finnane, Milton Lewis and Stephen Garton, whose studies 'Asylum, Families and the State' (1985), *Managing Madness: Psychiatry and Society in Australia 1788–1980* (1988) and *Medicine and Madness: A Social History of Insanity in New South Wales 1880–1940* (1988) placed institutionalised Australians within a framework informed by colonial histories of social control. Garton in particular applied a gendered perspective to the asylum patient, integrating feminist perspectives addressed by Jill Julius Matthews in her pioneering historical-sociological study *Good and Mad Women: The Historical Construction of Femininity in Twentieth-Century Australia* (1984). Works such as these established the historical principles for future historians both within Australia and Aotearoa New Zealand.

After a relatively quiet 1990s, where historians touched on psychiatry and mental health as a necessary consideration within the study of colonial lives (Goodman, 1994) or histories of war and conflict (Garton, 1996; Thomson, 1994), the beginning of the twenty-first century saw a renewed focus on mental health histories following on from the 'social turn' by a range of historians in both Australia and Aotearoa New Zealand. Prominent was a consideration of the institution itself, in particular its power over the institutionalised (Finnane, 2003) and its materiality (Brookes, 2011; Coleborne, 2001; Coleborne & MacKinnon, 2003,

2011; Monk, 2008). The colonial contexts in both Australia and Aotearoa New Zealand saw a re-examination of colonial history and eugenicist policies during this period, which integrated analyses of colonial mental health (Anderson, 2002; Bashford, 2003; Garton, 1994).

Much work on both Australia and Aotearoa New Zealand in the first 15 years of the twenty-first century was advanced by Australian historian Catharine Coleborne, who worked across the histories of both countries, along with scholars Barbara Brookes, Dolly MacKinnon, Lee-Ann Monk, Angela McCarthy and others. The transition from the institution to the gendered or racialised patient was an important shift during this period, signalling future studies that moved away from authorities and towards consumers of mental health care (Brookes & Thomson, 2001; Brickell, 2005; Burke & Coleborne, 2011; Coleborne, 2007, 2010, 2012, 2014, 2015; Dawson, 2012; Dunk, 2019; McCarthy, 2015; Paisley, 2009; Roth, 2018), as well as their families (Coleborne, 2006, 2010; Martyr, 2022).

The broadening of the 'social turn' has also more recently seen histories of mental health across both Australia and Aotearoa New Zealand more critically respond to current issues including worsening mental health crises, LGBTQIA+ health, climate change and military suicide. Key to the current mental health crises are the workings and justifications for deinstitutionalisation in the 1970s, prompted by increasing attention on social psychiatry, emerging in the 1950s and informed by the experiences of traumatised Second World War soldiers in the post-war period. The 2003 *Health and History* special issue entitled 'Histories of Psychiatry after Deinstitutionalisation: Australia and New Zealand' (MacKinnon & Coleborne, 2003) began the historical conversation about these themes. More recent studies have continued to analyse and present critical responses to the processes of deinstitutionalisation applied to the current state of community mental health care (Burge, 2015; Dunlop, 2022; Dunlop & Pols, 2022; Finnane, 2009; Rosen, 2022) and influenced by the rise of Mad Studies and consumer-led scholarship (Jeffs & Leggatt, 2020; O'Hagan, 2014).

At the time of writing, this shift is in progress and gaining momentum. While studies have largely followed trends in international scholarship, scholars in Australia and Aotearoa New Zealand have taken bold strides to incorporate the voices and choices of mental health consumers throughout the research process (Ellis & Coleborne, 2022). The Re;Minding Histories research group, based at the University of Sydney, is a co-produced project that actively resists hierarchical mental health research (see Chap. 5; https://www.remindinghistories.net/). Ongoing projects by research students are signalling the way forward, foregrounding lived experience and practitioner knowledge to produce truly critical histories, while recent histories by established scholars are demonstrating the necessity of criticality and interdisciplinarity to any historical study of mental health (Coleborne, 2020; Coleborne, 2024; Roberts-Pedersen, 2024).

Human impacts resulting from the global environmental crisis have informed the Planetary Health Histories research project (Anderson & Dunk, 2022; https://www.planetaryhealthhumanities.com/) and the Ecological Emotions Research Lab, run by psychologist Paul Rhodes and historian James Dunk (Rhodes & Dunk, 2023; https://www.eco-emotion.au/). The broadening recognition within History more broadly that studies examining the marginalisation of groups who have historically been pathologised must be led by members of those groups has extended into queer histories, incorporating transgender histories (Eames, 2019; Riseman, 2023). Although the number of Indigenous mental health studies, particularly those that are Indigenous-led, have increased across Australia and Aotearoa New Zealand (Bond et al., 2021; Kopua et al., 2020; Westerman & Sheridan, 2020), the way that psychiatry interacted with Indigenous lives in the colonial periods and beyond is a subject that requires far more in-depth investigation (see Chap. 7, Kopua et al.; Burke & Coleborne, 2011; Martyr & Davison, 2018; Watego et al., 2021).

The current mental health crisis in the Australian and Aotearoa New Zealand militaries, which in Australia has prompted a recent Royal Commission into Defence and Veteran Suicide, has transformed the existing interest in military mental health in these countries (Garton, 1996; Jenks & Wanhalla, 2020; Karageorgos, 2018, 2020; Roberts-Pedersen, 2014; Weaver & Wright, 2005) into a more critical and interdisciplinary exploration (see Chap. 2, Karageorgos; Karageorgos, 2023a, 2023b; Roberts-Pedersen, 2025; see also Australian Research Council

DP230101689 led by Ben Wadham, with Margaret Hutchison and Karen Bird, which aims to investigate the historical and social dimensions of veteran suicide).

Searching for a Central Thread in Sociology and Other Social Research

Perhaps distinct from historical approaches to critical mental health scholarship in this region is that attempting to write a succinct overview of social or even sociologically informed perspectives is more challenging. We have spent some time in this Introduction carefully outlining some of the development of History in this field to better integrate historical accounts within or alongside social accounts (see Chap. 10, Howe, for an example of this). In this section, we move to more cautiously share some insights into how 'the social' has emerged in relatively recent research in Australia and Aotearoa New Zealand. Here, we cite some useful examples but stress that this section likely excludes much of the work that is happening, albeit often isolated and fragmented from other colleagues and scholars in different institutions, disciplines and fields.[1]

(Post-)positivist paradigms of health and mental health are typically dominant within medicine, psychiatry, psychology, public health and other scientism-centric disciplines (Broom & Willis, 2007). Given this, we might then position *any* more socially attuned or interpretivist research to be an inherently critical perspective. This expands what we might label as critical mental health research to include studies of the experiences and discourses of mental health for different contexts, people, groups, communities and societies. Social researchers in the region have explored the experience and politics of psychiatric diagnosis (Hendry, 2020; Kokanović et al., 2006; Smith et al., 2015), narratives of mental ill-health or distress (Chowdhury et al., 2020), and representations and discourses of mental health (Holland, 2012; Horwood & Augoustinos, 2022). They have

[1] We also note Brossard and Chandler (2022), Cohen (2018b) and Harbusch (2022) as useful overviews and summaries of critical and social approaches to mental health within sociology and related disciplines, in Australia, Aotearoa New Zealand and internationally.

provided critical insights into how individuals and groups negotiate mental health alongside, for example, the ongoing impacts of colonisation (Coe, 2021; Cohen, 2021; Rhodes, 2020), experiences of racism or cultural and ethnic diversity (Gilbert et al., 2019; Killingsworth et al., 2010) or gender diversity and sexual diversity (Cover, 2005; Hectors, 2023; Howe, 2023; Jack, 2023; Lambert, 2023; Newman et al., 2021), and within particular contexts, including psychiatry, nursing and other clinical or care practices (Duff et al., 2020; Furler et al., 2010), work and employment (Ridge et al., 2019; Vella, 2023), policy and programming (O'Connor et al., 2023), and law and the legal system (Gooding et al., 2020; Maylea et al., 2021; Weller, 2012). This work has, importantly, engaged with critical issues of power, control, subjectivity and expertise related to mental health and psychiatry.

Other critical and social mental health scholars have continued to broaden their attention towards wellness practices and industries (Hendry, 2023; Smith et al., 2024), happiness psychology and economics (Hill et al., 2021), and therapeutic cultures (Wright, 2008). Much of this research also considers how digital media and technologies and digital cultures contribute to the expansion of psy-discourses and practices and the ever-evolving marketisation and commercialisation of wellbeing and mental health as digital products or e-health interventions. These efforts are often grounded in popular and clinical developments in neuroscience (see Chap. 8, Flore; Flore, 2023), social media platforms (see Chap. 11, Hendry; Hanckel, 2023) and bio-informatics (Gooding, 2019).

Theoretically and conceptually, many contributions draw on Mad Studies, critical psychiatry or psychology, and anti-psychiatry approaches (see Cohen, 2018b, for an overview, and Chap. 6, Sinclair and Chap. 3, Tseris). These scholars and activists offer critical perspectives that are central to their thinking, analysis and writing. This research challenges how we might understand mental health and invites us to rethink supposedly care-full service design, treatment and delivery in particular health contexts including nursing (Wand, 2024), social work (Macfarlane, 2020; Morley & Stenhouse, 2020; Tseris, 2019), psychiatry (Daya, Maylea, et al., 2020), counselling (McLeay et al., 2023), advocacy and service design (Daya, Hamilton, & Roper, 2020), and the psy-disciplines more broadly (Klein & Mills, 2017).

In Australia and Aotearoa New Zealand, this critical inquiry has also been motivated by psy-survivors and leadership and activism within the consumer/survivor/ex-patient movement and research exploring the peer work and lived experience sectors (Barlott et al., 2020; Byrne et al., 2018; Hurley et al., 2018; Sinclair et al., 2024). This has been crucial to teasing apart simple binary distinctions between service users or consumers and psychiatrists or other clinicians (Sinclair, Mahboub, et al., 2023). At the same time, we also see that lived experience and peer expertise and skills are increasingly co-opted into new forms of neoliberalism, psy-expansionism and organisational surveillance that maintain rather than challenge dominant assumptions about psychiatric expertise (see Chap. 6, Sinclair; Chap. 3, Tseris; Daya et al., 2020; Sinclair, Gillieatt, et al., 2023).

No matter how disconnected, this critical scholarship illustrates how normative concepts of 'mental health' are entrenched in policy and practice across Australia and Aotearoa New Zealand. Here, questioning the taken for granted has come to sit within the centre of how we understand critical mental health studies.

Chapter Overviews

While this book does not promise a complete overview of critical mental health scholarship and activity happening across Australia or Aotearoa New Zealand, we share these chapters as a glimpse of some of the vital and necessary critical work in this region. They provide critical interrogation of how personal, community, institutional and mediated relations make and remake experiences and knowledges of 'mental health'. Our intention is to not to have produced a handbook or a complete summary of critical historical or social scholarship in the field or the diverse research in our countries. Instead, we offer a snapshot of a point in time, where our contributors think alongside each other, without demanding one collective or shared perspective across the chapters. We note that not all contributors are directly engaged in psy-survivor work—like Voronka (2016) we too are cautious about essentialised notions of lived experience or expertise and make no demands for contributors to share or not share their own mental health experiences (see Chap. 6, Sinclair, for a critical reflection of care in relation to this), but we hope that this collection highlights that collective political and social ideas are important and underpin much of our thinking, writing and doing.

Some chapters focus specifically on the Aotearoa New Zealand or Australian context, and in particular the settler-colonial histories of these nation states. Illustrating this approach, Diana Kopua, Mark Kopua and Michelle Levy write about changing mental health systems in Aotearoa New Zealand through Indigenous ontology and epistemology that dismantles health inequity for Māori in Chap. 7. In Chap. 4, Roberto McLeay uses Foucauldian concepts to reconsider social and emotional learning policies in the Aotearoa New Zealand education system. Other chapters are examples of critical scholarship emerging from this region that attend to broader developments in relation to mental health or psychiatry. While these may not explicitly reference the authors' location, they are no doubt influenced by emerging trends or research in their region. For example, Jacinthe Flore's interrogation of the digital wearables that target the vagus nerve in Chap. 8 draws on histories of science and technology beyond Australia. Likewise, Kris Taylor writes in Chap. 9 about the 'making' of pornography addiction through popular media vernacular and contextualises his discussion with media examples from Aotearoa New Zealand and other media outlets from the Global North. In both accounts, we can see how psychiatry is a variously global project, one driven by developments towards diagnoses and treatment in the DSM-5 and the Global North more broadly. This international context nuances how psychiatry and psychology are understood and experienced locally and nationally.

Chapters 2 and 5 invite us to consider historical questions about how psychiatry, trauma and mental health are imagined and enacted in Australia. In Chap. 2, Effie Karageorgos analyses clinical understandings alongside the 'moral genealogy' of trauma posited by Fassin and Rechtman (2009) to describe why, despite increasing recognition of war-related trauma amongst soldiers and veterans, limited perspectives that focus on personal failings in the individual persist at the social and military levels. She follows the shifting importance of the 'traumatic memory' in assessments of war trauma from the nineteenth century to DSM-III (1980) to argue that despite a clinical shift away from heredity and personality and towards the traumatic event in the official assessment of post-traumatic stress disorder, submissions to the Australian Royal Commission into Defence and Veteran Suicide (2021–2024) reveal the continuation of outdated and harmful views on military mental health. The Re;Minding Histories Research Group presents their own pioneering co-produced historical research into Australian community psychiatry as a methodological case study in Chap. 5. This co-authored chapter contains frank and personal discussion by research leaders and investigators, which include academics as well as peer researchers and people with lived experience, about their role and experiences of a large-scale co-produced research project. By addressing the practical considerations required when embarking on such a project, including interview protocol, financial concerns and fairness in attribution and involvement, they provide a model for future scholarly investigations into the study of mental health that prioritise lived experience research.

Extending this attention to history, Chaps. 7 and 10 discuss how settler colonialism and colonial epistemologies have influenced how mental health systems and services enact care for different groups and communities in the past and, critically, the present. Diana Kopua, Mark Kopua and Michelle Levy detail the benefits of Mahi a Atua or a 'way of being', an Indigenous ancestral philosophy that prioritises Indigenous knowledge systems to address societal inequities in Chap. 7. Mahi a Atua combines Kuapapa Māori critical theory with systems change theory to transform systems using an Indigenous consciousness-based approach that integrates analysis of power, racism and privilege to benefit the lives of Indigenous people in Aotearoa. Their chapter details the path forward in addressing institutional racism causing inequities for Kuapapa Māori and signals the growing movement towards imagining productive and equitable Indigenous futures. Likewise, in Chap. 10, Rebecca Howe moves both forwards and backwards through time. Howe challenges the primacy of the progress narrative of depathologisation within trans health by creatively presenting eight international case studies concerning gender-affirming medical care that demonstrate the need to focus not only on time but also place when considering attitudes towards and practices of trans health. Howe's chapter demonstrates the variety of ways that state priorities have influenced the conceptualisation and justification of care in the present, focusing on the role of neoliberal, capitalist and colonial rationales and practices in trans health care, and ultimately arguing for the prominence of a rights-based discussion of the question of trans depathologisation.

In Chap. 6, Aimee Sinclair also questions the conditions of 'care' and power for psychiatric institutions in Australia and elsewhere. Sinclair's chapter engages with feminist new materialism and post-humanism to offer us a relational approach to care that champions a more life-affirming Mad care that actively attends to difference. Closely drawing on psy-survivor and Mad Studies work and activism, as well as her own and other peer support workers' experiences of the Australian mental health care system, Sinclair creatively outlines a way to 'think-feel-practice care differently'. Similarly, Emma Tseris in Chap. 3 queries how psy-professionals make decisions about how we understand distress, trauma and suffering, and invites us to consider how neoliberal and 'sanist' perspectives of mental health are deployed in the health system and higher education more broadly. Tseris reflects on her own experiences teaching and researching in social work and outlines the tensions she faces in trying to challenge dominant psychological and psychiatric knowledges. The chapter explains how more progressive perspectives—those related to lived experience or trauma-informed care—are often co-opted, dismissed or included in tokenistic ways that dampen their political power. Like Sinclair, Tseris calls for attention to the inequities that the psychiatric system perpetuates rather than alleviates.

Our remaining contributors turn their attention to discourses and applications of psychiatry, mental health and wellbeing in applied contexts: psychology, education, digital mental health and social media. Writing about how 'pornography addiction' is deployed in popular media and wider social vernacular, Kris Taylor, in Chap. 9, explores how this informal diagnosis exists within 'a peculiar space'. Porn addiction is not included in the DSM-5 but is made 'real' outside official categories where the concept and term come to stand in for any and all problems with pornography viewing and its presumed 'effects'. Taylor illustrates how this both challenges and extends formal psychiatric diagnosis classification. In Chap. 4, Roberto McLeay analyses how early childhood education interventions and policies related to social and emotional learning and positive psychology inspire children and their communities to govern their own emotional behaviour and that of others. Using Foucault's concepts of governmentality and pastoral power, McLeay demonstrates how the national early childhood curriculum in Aotearoa New Zealand, *Te Whāriki* and other supplementary resources encourage teachers to surveil and direct children's emotional competence in ways that direct them 'towards their salvation'. This chapter stresses how seemingly benevolent psy-discourses in schools subtly urge students to govern themselves in increasingly constrained and individual ways.

Jacinthe Flore expands this attention to disciplining bodies and emotions in Chap. 8 and considers how digitally tracking and datafying the vagus nerve (the central nerve of the parasympathetic nervous system) have become positioned by wellbeing technology companies as more 'holistic' approaches to mental health. Her chapter reflects the Australian context where we see increasing attention towards seemingly benign and benevolent neuroscience interventions in mental health and a widening local commercial market for 'objective' digital interventions and cost-effective 'care'. Engaging with new materialist approaches, Flore argues that the vagus nerve and its influence on the gut have become the 'brain chemicals' for the wellbeing and mental health industry. Also exploring digital media and mental health, in Chap. 11, Natalie Ann Hendry interrogates how the psychiatric concept and phenomenon of 'social contagion' has been translated for the digital age into digital or social media contagion. This frames young people's engagement with digital media—the content they scroll through, the connections they make—as always already at risk of being highly suggestible, self-diagnosing mental disorders and illnesses. Hendry argues that 'contagion' simplifies and ignores the complexity of both media and mental health, and suggests that this is important for clinical psychiatry research as much as it is an important consideration for critical mental health scholars.

Our collection ends with an epilogue from the series editor, Bruce M.Z. Cohen, a supporter of the Social Production of Mental Health seminar series since its first year. He points to the rise and importance of Mad Studies scholarship and how, despite their interdisciplinarity, the critical mental health communities in Australia, Aotearoa New Zealand

and elsewhere are united by their parallel concerns about the place of mental health care within capitalist systems.

Our Hope for This Book

There are, no doubt, important perspectives and approaches that are missing from this collection. We trust that other colleagues and groups doing similarly critical work will find the collection affirming and encouraging, and that their own critical perspectives find kindred, albeit loose and diverse, community alongside these chapters and the ideas they share. We are eager to witness how the chapters are read, shared and used as springboards for other critical inquiries and, of course, how readers extend and challenge these perspectives in ways that continue to develop critical mental health scholarship in Australia and Aotearoa New Zealand, and internationally.

We acknowledge, too, that not all of the planned contributions to this book were included as the collection progressed. This is not due to any chapters being misaligned to the book's aims, but rather that critical mental health scholarship is often sidelined within and beyond the academy. Critical inquiry and research in the field of mental health can at times be marginal to our other institutional roles, especially where we may be isolated from others with related intentions. For some authors, more immediate commitments to their lives, families or communities, or the demands of hope-affirming activism or precarious employment meant that they could not participate. This reminds us of the political and material conditions our work; we hope that this collection encourages other researchers that this work is valid and crucial, even with its likely limitations and any unintentional omissions. We genuinely thank the contributors whose work is not in this first collection, as their ideas, fieldwork and wisdom have shaped this book and our thinking about critical mental health.

We hope this book finds a readership that appreciates the value of critical inquiry *and* is also challenged by the complexity and scope of this work. Putting aside that critical perspectives are often marginalised or how fragmented the disciplines in this field are, what can be more socially useful than critical thinking? What is more crucial for ethical citizenship, communities, researchers and practitioners than to look carefully at and ask questions about mental health in all its forms and developments in Australia and Aotearoa New Zealand? What is more necessary than asking more critical and contextually engaged questions, rather than asking questions that merely reinforce the status quo? We hope this collection goes, at least part of the way, towards those critical questions and thinking of—and remaking too—a different world.

References

American Psychiatric Association. (2013). *Diagnostic and statistical manual of mental disorders* (5th ed.). American Psychiatric Association.

Anderson, W. (2002). *The cultivation of whiteness: Science, health and racial destiny in Australia*. Melbourne University Press.

Anderson, W., & Dunk, J. (2022). Planetary health histories: Toward new ecologies of epidemiology? *Isis, 113*(4), 767–788. https://doi.org/10.1086/722308

Barlott, T., Shevellar, L., Turpin, M., & Setchell, J. (2020). Destabilising social inclusion and recovery, and pursuing "lines of flight" in the mental health sector. *Sociology of Health & Illness, 42*(6), 1328–1343. https://doi.org/10.1111/1467-9566.13106

Bashford, A. (2003). *Imperial hygiene: A critical history of colonialism, nationalism and public health*. Springer.

Bond, C. J., Singh, D., & Tyson, S. (2021). Black bodies and bioethics: Debunking mythologies of benevolence and beneficence in contemporary Indigenous health research in colonial Australia. *Journal of Bioethical Inquiry, 18*(1), 83–92. https://doi.org/10.1007/s11673-020-10079-8

Brickell, C. (2005). Same-sex desire and the asylum: A colonial experience. *New Zealand Journal of History, 39*(2), 158–178.

Brookes, B. (2011). The asylum lens: Photographs in the Seacliff Asylum case files. In A. Wanhalla & E. Wolf (Eds.), *Early New Zealand photography: Images and essays* (pp. 98–103). Otago University Press.

Brookes, B., & Thomson, J. (Eds.). (2001). *'Unfortunate folk': Essays on mental health treatment 1863–1992*. Otago University Press.

Broom, A., & Willis, E. (2007). Competing paradigms and health research. In M. Saks & J. Allsop (Eds.), *Researching health: Qualitative, quantitative and mixed methods* (pp. 16–31). Sage.

Brossard, B., & Chandler, A. (2022). *Explaining mental illness: Sociological perspectives*. Bristol University Press.

Burge, R. (2015). Callan Park in transition. In D. Kritsotaki, V. Long, & M. Smith (Eds.), *Deinstitutionalisation and after: Post-war psychiatry in the Western world* (pp. 57–74). Palgrave.

Burke, L., & Coleborne, C. (2011). Insanity and ethnicity in New Zealand: Māori encounters with the Auckland Mental Hospital, 1860–1900. *History of Psychiatry, 22*(3), 285–301. https://doi.org/10.1177/0957154x10390435

Byrne, L., Roennfeldt, H., O'Shea, P., & Macdonald, F. (2018). Taking a gamble for high rewards? Management perspectives on the value of mental health peer workers. *International Journal of Environmental Research and Public Health, 15*(4), Article 4. https://doi.org/10.3390/ijerph15040746

Chowdhury, N., Gibson, K., & Wetherell, M. (2020). Polyphonies of depression: The relationship between voices-of-the-self in young professional women aka 'top girls'. *Health, 24*(6), 773–790. https://doi.org/10.1177/1363459319846934

Coe, G. (2021). Decolonising mental illness: Aboriginal and Torres Strait Islander peoples' Social and Emotional Wellbeing and the persistence of colonial discourses in policy. *Australian Journal of Social Issues, 56*(4), 503–521.

Cohen, B. M. Z. (2018a). Introduction: The importance of critical approaches to mental health and illness. In B. M. Z. Cohen (Ed.), *Routledge international handbook of critical mental health* (pp. 1–12). Routledge.

Cohen, B. M. Z. (Ed.). (2018b). *Routledge international handbook of critical mental health*. Routledge.

Cohen, B. M. Z. (2021). A postcolonial critique of mental health: Empire and psychiatric expansionism. In R. Moodley & E. Lee (Eds.), *The Routledge international handbook of race, culture and mental health* (pp. 32–42). Routledge.

Coleborne, C. (2001). Exhibiting "madness": Material culture and the asylum. *Health and History, 3*(2), 104–117. https://doi.org/10.2307/40111408

Coleborne, C. (2006). 'His brain was wrong, his mind astray': Families and the language of insanity in New South Wales, Queensland and New Zealand, 1880s–1910. *Journal of Family History, 31*(1), 45–65. https://doi.org/10.1177/0363199005283009

Coleborne, C. (2007). *Reading 'madness': Gender and difference in the colonial asylum in Victoria, Australia, 1848–1888*. Network Books, Australian Public Intellectual Network.
Coleborne, C. (2010). *Madness in the family: Insanity and institutions in the Australasian colonial world, 1860–1914*. Palgrave Macmillan.
Coleborne, C. (2012). Insanity, gender and empire: Women living a 'loose kind of life' on the colonial institutional margins, 1870–1910'. *Health and History, 14*(1), 77–99. https://doi.org/10.1353/hah.2012.0030
Coleborne, C. (2014). White men and weak masculinity: Men in the public asylums in Victoria and New Zealand, 1860s–1900s. *History of Psychiatry, 25*(4), 468–476. https://doi.org/10.1177/0957154x14543758
Coleborne, C. (2015). *Insanity, identity and empire: Colonial institutional confinement in Australia and New Zealand, 1870–1910*. Manchester University Press.
Coleborne, C. (2020). *Why talk about madness? Bringing history into the conversation*. Springer Nature.
Coleborne, C. (2024). Vagrant Lives in Colonial Australasia: *Regulating Mobility* 1840–1910. Bloomsbury.
Coleborne, C., & MacKinnon, D. (Eds.). (2003). *'Madness' in Australia: Histories, heritage and the asylum*. University of Queensland Press.
Coleborne, C., & Mackinnon, D. (Eds.). (2011). *Exhibiting madness in museums: Remembering psychiatry through collections and display*. Routledge.
Cover, R. (2005). Queer subjects of suicide: Cultural studies, sexuality and youth suicide concepts in New Zealand. *New Zealand Sociology, 20*(1), 78–101.
Dawson, M. (2012). Halting the 'sad degenerationist parade': Medical concerns about heredity and racial degeneracy in New Zealand psychiatry, 1853–1899. *Health and History, 14*(1), 38–55. https://doi.org/10.1353/hah.2012.0013
Daya, I., Hamilton, B., & Roper, C. (2020). Authentic engagement: A conceptual model for welcoming diverse and challenging consumer and survivor views in mental health research, policy, and practice. *International Journal of Mental Health Nursing, 29*(2), 299–311. https://doi.org/10.1111/inm.12653
Daya, I., Maylea, C., Raven, M., Hamilton, B., & Jureidini, J. (2020). Defensive rhetoric in psychiatry: An obstacle to health and human rights. *The Lancet Psychiatry, 7*(3), 231. https://doi.org/10.1016/S2215-0366(20)30045-6
Duff, C., Kokanović, R., Flore, J., Thomas, S. D. M., Callard, F., & Blackman, L. (2020). Perspectives on person-centred care for borderline personality disorder: A critical research agenda. *Health Sociology Review, 29*(1), 1–15. https://doi.org/10.1080/14461242.2020.1715815

Dunk, J. (2019). *Bedlam at Botany Bay*. NewSouth.

Dunlop, R. (2022). Disaster responses and Commonwealth involvement in community mental health, 1973–1975. *History Australia, 19*(1), 115–132. https://doi.org/10.1080/14490854.2022.2028555

Dunlop, R., & Pols, H. (2022). Deinstitutionalisation and mental health activism in Australia: Emerging voices of individuals with lived experience of severe mental distress, 1975–1985. *History Australia, 19*(1), 92–114. https://doi.org/10.1080/14490854.2022.2028559

Eames, R. (2019). Problem bodies and queer legacies: Rethinking approaches to trans history in the case of Harry Crawford, Sydney, 1920. *Lilith: A Feminist History Journal, 25*, 50–62.

Ellis, R., & Coleborne, C. (2022). Co-producing madness: international perspectives on the public histories of mental illness. *History Australia, 19*(1), 133–150. https://doi.org/10.1080/14490854.2022.2028558

Fassin, D., & Rechtman, R. (2009). *The empire of trauma: An inquiry into the condition of victimhood*. Princeton University Press.

Finnane, M. (1985). Asylums, families and the state. *History Workshop, 20*, 134–148. https://doi.org/10.1093/hwj/20.1.134

Finnane, M. (2003). The ruly and the unruly: Isolation and inclusion in the management of the insane. In C. Strange & A. Bashford (Eds.), *Isolation: Places and practices of exclusion* (pp. 89–103). Routledge.

Finnane, M. (2009). Opening up and closing down: Notes on the end of an asylum. *Health and History, 11*(1), 9–24. https://doi.org/10.1353/hah.2009.0031

Flore, J. (2023). *The artefacts of digital mental health*. Springer.

Foucault, M. (1964). *Madness and civilization: A history of insanity in the age of reason*. Vintage.

Furler, J., Kokanović, R., Dowrick, C., Newton, D., Gunn, J., & May, C. (2010). Managing depression among ethnic communities: A qualitative study. *The Annals of Family Medicine, 8*(3), 231–236. https://doi.org/10.1370/afm.1091

Garton, S. (1988). *Medicine and madness: A social history of insanity in New South Wales 1880–1940*. UNSW Press.

Garton, S. (1994). Sound minds and healthy bodies: Re-considering eugenics in Australia, 1914–1940. *Australian Historical Studies, 26*(103), 163–181. https://doi.org/10.1080/10314619408595958

Garton, S. (1996). *The cost of war: Australians return*. Oxford University Press.

Gilbert, A. S., Antoniades, J., Bowen, Z., & Brijnath, B. (2019). Legitimising depression: community perspectives and the help-seeking continuum. *Health*

Sociology Review, 28(3), 291–306. https://doi.org/10.1080/14461242.2019.1670090
Gooding, P. (2019). Mapping the rise of digital mental health technologies: Emerging issues for law and society. *International Journal of Law and Psychiatry, 67*, 101498. https://doi.org/10.1016/j.ijlp.2019.101498
Gooding, P., McSherry, B., & Roper, C. (2020). Preventing and reducing 'coercion' in mental health services: An international scoping review of English-language studies. *Acta Psychiatrica Scandinavica, 142*(1), 27–39. https://doi.org/10.1111/acps.13152
Goodman, D. (1994). *Gold seeking: Victoria and California in the 1850s*. Allen & Unwin.
Hanckel, B. (2023). *LGBT+ youth and emerging technologies in Southeast Asia: Designing for wellbeing*. Springer.
Harbusch, M. (Ed.). (2022). *Troubled persons industries: The expansion of psychiatric categories beyond psychiatry*. Springer Nature.
Hectors, A. (2023). Homosexuality in the DSM: A critique of depathologisation and heteronormativity. *New Zealand Sociology, 38*(1), 18–28.
Hendry, N. A. (2020). Young women's mental illness and (in-)visible social media practices of control and emotional recognition. *Social Media + Society, 6*(4), 1–10. https://doi.org/10.1177/2056305120963832
Hendry, N. A. (2023). "Hey lovely! Don't miss this opportunity!": Digital temporalities of wellness culture, email marketing and the promise of bodily abundance. *Journal of Sociology, 59*(3), 664–681. https://doi.org/10.1177/14407833221101397
Hill, N., Brinkmann, S., & Petersen, A. (Eds.). (2021). *Critical happiness studies*. Routledge.
Holland, K. (2012). The unintended consequences of campaigns designed to challenge stigmatising representations of mental illness in the media. *Social Semiotics, 22*(3), 217–236. https://doi.org/10.1080/10350330.2011.648398
Horwood, G., & Augoustinos, M. (2022). 'It's more than sadness': The discursive construction of depression on Australian depression websites. *Qualitative Health Research, 32*(7), 1185–1196. https://doi.org/10.1177/10497323221102240
Howe, R. (2023). Unmapping social work scholarship about gender self-designation: Reconstructing the basis for engagement. *Critical and Radical Social Work, 11*(2), 279–294. https://doi.org/10.1332/204986021X16521784495123

Hurley, J., Cashin, A., Mills, J., Hutchinson, M., Kozlowski, D., & Graham, I. (2018). Qualitative study of peer workers within the 'Partners in Recovery' programme in regional Australia. *International Journal of Mental Health Nursing, 27*(1), 187–195. https://doi.org/10.1111/inm.12308

Jack, K. (2023). A queer critique of psychiatric knowledge: Medicalising queer sadness. *New Zealand Sociology, 38*(1), 81–94.

Jeffs, S., & Leggatt, M. (2020). *Out of the madhouse: From asylums to caring community?* Australian Scholarly Publishing.

Jenks, T., & Wanhalla, A. (2020). Psychological casualties: War neurosis, rehabilitation, and the family in post-World War II New Zealand. *Health & History, 22*(2), 1–25. https://doi.org/10.5401/healthhist.22.2.0001

Karageorgos, E. (2018). Mental illness, masculinity, and the Australian soldier: Military psychiatry from South Africa to the First World War'. *Health and History, 20*(2), 10–29. https://doi.org/10.1353/hah.2018.0015

Karageorgos, E. (2020). The bushman at war: Gendered medical responses to combat breakdown in South Africa, 1899–1902. *Journal of Australian Studies, 44*(1), 18–32. https://doi.org/10.1080/14443058.2020.1717581

Karageorgos, E. (2023a). Medical fears of the malingering soldier: 'Phony cronies' and the Repat in 1960s Australia. *Medical History, 67*(2), 172–191. https://doi.org/10.1017/mdh.2023.19

Karageorgos, E. (2023b). 'The unseen enemy persists': Delusion, trauma and the South African War in Australian asylum case notes. *Social History of Medicine, 36*(2), 316–336. https://doi.org/10.1093/shm/hkac049

Killingsworth, B., Kokanović, R., Tran, H., & Dowrick, C. (2010). A care-full diagnosis: Three Vietnamese-Australian women and their experience of becoming "mentally ill". *Medical Anthropology Quarterly, 4*(1), 108–123. https://doi.org/10.1111/j.1548-1387.2010.01087.x

Klein, E., & Mills, C. (2017). Psy-expertise, therapeutic culture and the politics of the personal in development. *Third World Quarterly, 38*(9), 1990–2008. https://doi.org/10.1080/01436597.2017.1319277

Kokanović, R., Petersen, A., & Klimidis, S. (2006). 'Nobody can help me … I am living through it alone': Experiences of caring for people diagnosed with mental illness in ethno-cultural and linguistic minority communities. *Journal of Immigrant Health, 8*, 125–135. https://doi.org/10.1007/s10903-006-8521-0

Kopua, D., Kopua, M. A., & Bracken, P. J. (2020). Mahi a Atua: A Māori approach to mental health. *Transcultural Psychiatry, 57*(2), 375–383. https://doi.org/10.1177/1363461519851606

Laing, R. D. (1965). *The divided self: An existential study in sanity and madness*. Penguin.
Lambert, V. (2023). Gender dysphoria and the medicalisation of distress. *New Zealand Sociology, 38*(1), 56–67.
Lewis, M. (1988). *Managing madness: Psychiatry and society in Australia 1788–1980*. AGPS.
Macfarlane, S. (2020). Opening spaces for alternative understandings in mental health practice. In B. Pease, J. Allen, & L. Briskman (Eds.), *Critical social work* (2nd ed., pp. 201–213). Routledge.
MacKinnon, D., & Coleborne, C. (Eds.). (2003). *Health and History* (Special issue: Deinstitutionalisation), *5*(2).
Martyr, P. (2022). A town without pity? Three stories of public exposure, print media, and family histories of madness in Western Australia. *History Australia, 19*(1), 54–72. https://doi.org/10.1080/14490854.2022.2028554
Martyr, P., & Davison, S. (2018). Aboriginal people in Western Australian mental hospitals, 1903–1966. *Social History of Medicine, 31*(3), 462–484. https://doi.org/10.1093/shm/hkx006
Matthews, J. J. (1984). *Good and mad women: The historical construction of femininity in twentieth-century Australia*. Allen & Unwin.
Maylea, C., Katterl, S., Johnson, B., Alvarez-Vasquez, S., Hill, N., & Weller, P. (2021). Consumers' experiences of rights-based mental health laws: Lessons from Victoria, Australia. *International Journal of Law and Psychiatry, 78*, 101737. https://doi.org/10.1016/j.ijlp.2021.101737
McCarthy, A. (2015). *Migration, ethnicity, and madness: New Zealand, 1860–1910*. Otago University Press.
McLeay, R., Powell, D., & Cohen, B. M. Z. (2023). Young people's voices on emotions: A narrative inquiry. *International Journal of Qualitative Studies in Education, 1–16*. https://doi.org/10.1080/09518398.2023.2233945
Monk, L. (2008). *Attending madness: At work in the Australian colonial asylum*. Rodopi.
Morley, C., & Stenhouse, K. (2020). Educating for critical social work practice in mental health. *Social Work Education, 40*(1), 80–94. https://doi.org/10.1080/02615479.2020.1774535
Newman, C. E., Smith, A. K. J., Duck-Chong, E., Vivienne, S., Davies, C., Robinson, K. H., & Aggleton, P. (2021). Waiting to be seen: Social perspectives on trans health. *Health Sociology Review, 30*(1), 1–8. https://doi.org/10.1080/14461242.2020.1868900

O'Connor, C. A., Zufferey, C., & de Anstiss, H. (2023). Governing failed neoliberal subjects: Representations of women's mental health in Australian mental health policies. *Qualitative Social Work*. https://doi.org/10.1177/14733250231222020

O'Hagan, M. (2014). *Madness made me: A memoir*. Open Box.

Paisley, F. (2009). Peeling back history: The remembering Goodna exhibition. *Health and History (Special Issue: Australian Asylums and Their Histories), 11*(1), 172–178. https://doi.org/10.1353/hah.2009.0016

Rhodes, L. (2020). The colonising effect of western mental health discourses. *Social Work & Policy Studies: Social Justice, Practice and Theory, 2*(2), 1–16.

Rhodes, P., & Dunk, J. (2023). Eco-psychology: A critical paradigm in the climate emergency. *Australian Psychologist, 58*(3), 154–160. https://doi.org/10.1080/00050067.2022.2157240

Ridge, D., Broom, A., Kokanović, R., Ziebland, S., & Hill, N. (2019). Depression at work, authenticity in question: Experiencing, concealing and revealing. *Health, 23*(3), 344–361. https://doi.org/10.1177/1363459317739437

Riseman, N. (2023). *Transgender Australia: A History Since 1910*. Melbourne University Press.

Roberts-Pedersen, E. (2014). Damage: The 'war neurotic' serviceman comes home. *History Australia, 11*(1), 82–101. https://doi.org/10.1080/14490854.2014.11668501

Roberts-Pedersen, E. (2024). *Making mental health: A critical history*. Routledge.

Roberts-Pedersen, E. (forthcoming, 2025). *Human salvage: Suffering, adaptation and survival in the Second World War*. Cambridge University Press.

Rosen, A. (2022). From aftermath psychiatry to lattermath psychiatry, or from deinstitutionalisation to community-centric mental health services, and now to mental health care ecosystems: Beyond symbols, stereotypes, and stigma of home visiting and assertive community teams. *Health and History, 24*(2), 161–168. https://doi.org/10.1353/hah.2022.0023

Roth, D. (2018). The 'Red Plague' and general paralysis of the insane at Callan Park Hospital for the Insane in New South Wales 1877–1920. *Health and History, 24*(1), 109–127. https://doi.org/10.1353/hah.2022.0005

Scheff, T. J. (1963). The role of the mentally ill and the dynamics of mental disorder: A research framework. *Sociometry, 25*(4), 436–453. https://doi.org/10.2307/2786147

Sinclair, A., Fernandes, C., Gillieatt, S., & Mahboub, L. (2024). Peer work in Australian mental health policy: What 'problems' are we solving and to what effect(s)? *Disability & Society, 39*(7), 1656–1681. https://doi.org/10.1080/09687599.2022.2160926

Sinclair, A., Gillieatt, S., Fernandes, C., & Mahboub, L. (2023). Inclusion as assimilation, integration, or co-optation? A post-structural analysis of inclusion as produced through mental health research on peer support. *Qualitative Health Research, 33*(6), 543–555. https://doi.org/10.1177/10497323231163735

Sinclair, A., Mahboub, L., Gillieatt, S., & Fernandes, C. (2023). 'You just treat me like a human being': Using lived experience to (re)imagine boundary practices in mental health settings. *The British Journal of Social Work, 53*(3), 1408–1425. https://doi.org/10.1093/bjsw/bcad044

Smith, L., Hill, N., & Kokanović, R. (2015). Experiences of depression, the role of social support and its impact on health outcomes. *Journal of Mental Health, 24*(6), 342–346. https://doi.org/10.3109/09638237.2014.954693

Smith, N., Southerton, C., & Clark, M. (Eds.). (2024). *Researching contemporary wellness cultures*. Emerald.

Szasz, T. S. (1974). *The myth of mental illness: Foundations of a theory of personal conduct*. Harper & Row.

Thomson, A. (1994). *Anzac memories: Living with the legend*. Oxford University Press.

Tseris, E. (2019). Social work and Women's mental health: Does trauma theory provide a useful framework? *The British Journal of Social Work, 49*(3), 686–703. https://doi.org/10.1093/bjsw/bcy090

Vella, S. (2023). Community psychiatry and the medicalisation of unemployment. *New Zealand Sociology, 38*(1), 6–17.

Voronka, J. (2016). The Politics of 'people with lived experience': Experiential authority and the risks of strategic essentialism. *Philosophy, Psychiatry, & Psychology, 23*(3–4), 189–201. https://doi.org/10.1353/ppp.2016.0017

Wand, T. (2024). We have to cancel psychiatric nursing and forge a new way forward. *International Journal of Mental Health Nursing, 33*(2), 215–219. https://doi.org/10.1111/inm.13301

Watego, C., Whop, L. J., Singh, D., Mukandi, B., Macoun, A., Newhouse, G., Drummond, A., Mcquire, A., Stajic, J., Kajlich, H., & Brough, M. (2021). Black to the future: Making the case for Indigenist health humanities. *International Journal of Environmental Research and Public Health, 18*(16), 1–10. https://doi.org/10.3390/ijerph18168704

Weaver, J., & Wright, D. (2005). Shell shock and the politics of asylum committal in New Zealand, 1916–1922. *Health and History, 7*(1), 17–40. https://doi.org/10.2307/40111512

Weller, P. (2012). *New law and ethics in mental health advance directives: The Convention on the Rights of Persons with Disabilities and the right to choose.* Routledge.

Westerman, T. G., & Sheridan, L. (2020). A whole of community Indigenous suicide intervention program. Results from high risk remote communities. *Australian Psychologist, 55*, 363–374. https://doi.org/10.1111/ap.12470

Wright, K. (2008). Theorizing therapeutic culture: Past influences, future directions. *Journal of Sociology, 44*(4), 321–336. https://doi.org/10.1177/1440783308097124

Part I

Contesting Psy-Knowledges in Institutional Spaces

2

Locating the 'moral genealogy' of War Trauma in the Australian Royal Commission into Defence and Veteran Suicide

Effie Karageorgos

Introduction

On 8 July 2021 the Royal Commission into Defence and Veteran Suicide was established to investigate 'systemic issues and any common themes', as well as 'contributing risk factors relevant to defence and veteran deaths by suicide, including the possible contribution of pre-service, service (including training and deployments), transition, separation and post-service issues'. It aimed to focus on the 'impact of culture within the ADF [Australian Defence Force], Department of Defence and the Department of Veterans' Affairs [DVA]' as well as the 'availability, accessibility, timeliness and quality of health, wellbeing and support services' (Royal Commission into Defence and Veteran Suicide, 2021). The Royal Commission received 5889 individual or group submissions in total,

E. Karageorgos (✉)
School of Humanities, Creative Industries and Social Sciences,
University of Newcastle, Newcastle, NSW, Australia
e-mail: effie.karageorgos@newcastle.edu.au

N. A. Hendry, E. Karageorgos (eds.), *Critical Mental Health in Australia and Aotearoa New Zealand*, The Politics of Mental Health and Illness,
https://doi.org/10.1007/978-3-031-82861-4_2

1281 of which are published online. Many of these report negative interactions with ADF or DVA authorities in relation to physical or mental health issues. One prominent theme in these submissions is the unsuitability of military culture to the provision of adequate mental health care. These reports appear to contradict the ADF aim to ensure that all personnel are 'Fit to Fight, Fit for Work, Fit for Life', as affirmed in the Defence Mental Health and Wellbeing Strategy 2018–2023 (DMHWS) (Department of Defence, 2017). Many submissions claim that general attitudes towards psychological issues among some military superiors still associate mental illness with weakness or shame, resulting in reluctance to report by personnel, and subsequent worsening of any difficulties (Royal Commission into Defence and Veteran Suicide, 2022, pp. 36, 37). Again, this does not accord with the DMHWS contention that 'seeking help early and engaging in effective treatment can lead to improved outcomes and prevent future problems' (Department of Defence, 2017, p. 14). The Royal Commission's final report was presented on 9 September 2024, which, along with the submissions, suggests that a dramatic shift beyond the realm of military culture will be necessary to effectively support personnel throughout their period of service and beyond.

The findings of the Royal Commission also fail to reflect both clinical and cultural developments in mental health care. Connections made between psychiatric issues and personal attributes in the submissions and interim report instead echo both official and unofficial historical approaches towards war-related trauma within the Australian military forces throughout the twentieth century, which focused on deficiencies in the individual experiencing traumatic symptoms rather than the event or circumstance that precipitated those symptoms. Clinical developments in military psychiatry officially moved away from personality or heredity in the diagnosis of trauma upon the introduction of post-traumatic stress disorder (PTSD) in the Third Edition of the *Diagnostic and Statistical Manual of Mental Disorders* (DSM-III) of 1980. The neo-Kraepelinian approach within DSM-III offered the first standardised psychiatric nosology, meaning that a specific set of criteria was required for psychiatrists to recommend a diagnosis of PTSD, with its 'essential feature … the development of characteristic symptoms following a psychologically traumatic event that is generally outside the range of usual human experience'

(APA, 1980, p. 236). This important shift, instigated by American psychiatrists Chaim Shatan and Robert Lifton, legitimised the traumatic conditions of thousands of Vietnam veterans globally and promised to make them officially pensionable with no risk to reputation. When placed in historical context, it is evident that the clinical recognition of military trauma has advanced beyond earlier understandings, but this is not reflected in the findings of the current Royal Commission.

The soldier still occupies a particular space in the Australian imaginary beyond the ADF that is in need of radical revision. The character of the 'imaginary Australian', posited by Dixson, is determined by not only national priorities and individual personality or attributes but also 'socio-historical circumstance', which includes class, gender and ethnicity (Dixson, 1999, pp. 159–160; Anderson, 1983, p. 6). The identity of the soldier or veteran is thus predetermined by a combination of expectations set by and accepted within Australian society, and to which military personnel are assumed to adhere. Experiences which do not accord with these expectations, which in the Australian context are bound up in Anzac mythology emerging in the First World War (WWI) period, have been frequently minimised or questioned by military authorities, members of society and medical professionals (Thomson, 2013, p. 247; Karageorgos, 2023).

Fassin and Rechtman (2009, pp. 6, 153) contend that the 'moral genealogy' of trauma has historically appeared as 'a series of appropriations and dispossessions', with individuals included or excluded disproportionately based on criteria beyond clinical definitions. Their conceptualisation of the 'moral genealogy' draws on Foucault's genealogical analysis of the ways that a '"regime" of truth' is produced and regulated within institutions and societies, which positions truth as connected to 'systems of power which produce and sustain it' (Foucault, 1977, p. 14). Although the 'moral genealogy' of trauma does exist alongside the clinical, it gains its power from the 'collective process by which society defines its values and norms, and embodies them in individual subjects', which is both hierarchical and inequitable, meaning that some are accepted as victims and others are not (Fassin & Rechtman, 2009, p. 30). This focus on the 'moral' is not specific to the military context. In his Marxist approach to mental illness, sociologist Bruce M.Z. Cohen (2016, p. 33) writes that 'moral judgements' made by psy-professionals about 'deviants' are a form of social control that serves the broader system, an idea that reflects the broad scholarship on social influences on mental health care (see also Braslow, 1997; Rose, 2019; Scull, 2019). Arguably, the focus on morality is especially applicable to soldiers and veterans due to the sacred place they occupy within Australian nationalism, and the way this has subsequently influenced their consideration within psychiatry (Holbrook, 2014; Lake & Reynolds, 2010). Australian military identity, then, can be viewed as one of Foucault's 'truths'—one devised in the early decades of the twentieth century to project an image of a strong nation, and one which excludes the figure of the traumatised soldier or veteran.

In examining the moral genealogy of trauma, Fassin and Rechtman (2009, p. 5) assert that they 'seek to grasp the historic moment when suspicion ended', or when 'proof' replaced doubt as the primary response to traumatic symptoms. The focus away from individual psychology and towards responses to the traumatic event within the categorisation of PTSD meant that traumatised Australian soldiers were technically able to claim victim status rather than being viewed as 'slightly suspect individuals' (Twomey, 2013, pp. 104–105). This chapter argues that despite a shift in official ADF and DVA attitudes towards psychiatry, which reflect clinical developments from the late twentieth century about PTSD and associated conditions, 'suspicion' has not yet 'ended' in the Australian context. Although the ADF has taken unprecedented steps towards acknowledging the existence and prevalence of trauma among military personnel and has attempted to transform ADF culture in this respect, these efforts are futile without a parallel transformation of social views. Fassin and Rechtman (2009, p. 274) conclude that the 'truth of trauma lies not in psyche, mind or brain but in the moral economy of contemporary societies'. Placing publicly available submissions to the current Royal Commission into Defence and Veteran Suicide about military approaches to soldier and veteran mental health within their historical context highlights the continuing prominence of the moral genealogy of war trauma over the clinical. Analysing current responses to war trauma alongside its historical context suggests that the solution to the problem emphasised within the Royal Commission lies beyond the ADF or even the Department of Defence; it highlights the need for a change in broader accepted attitudes towards the soldier in Australian society.

The Traumatic Memory to the First World War

Collective trauma is produced and has varied meaning depending on its setting, as 'trauma scripts are performed in the theatres of everyday collective life' (Alexander, 2012, p. 11). Definitions of trauma existing within and emerging from the military setting are simply 'scripts' specific to that setting. This means that despite shifts in the clinical understandings of trauma, individuals within the ADF have continued to draw on their own culture and knowledges to shape psychiatric issues affecting Australian military personnel. Examining the crucial traumatic memory within the conceptualisation of war trauma from the nineteenth century until its resurgence in DSM-III allows us to chart the failure of the 'moral genealogy' of war trauma to move past pre-DSM-III conceptions of military mental health. Viewing the traumatic memory as a 'manmade object' draws attention to the ways its changing definition has reflected military priorities (Young, 1995, p. 141). Specifically, this has included a lingering preoccupation with heredity and personality in the assessment of trauma, which reached its peak from the Second World War years.

The dual genealogy posited by Fassin and Rechtman and others, which sees separate yet parallel developments in the scientific and moral recognition of trauma, has its roots in the European study of the 'psychic trauma' from the late nineteenth century, and specifically in the work of Jean-Martin Charcot, Pierre Janet and Sigmund Freud (Fassin & Rechtman, 2009, p. 29). The British surgeon John Erichsen first described the mechanism which would eventually lead to the conceptualisation of PTSD in the late twentieth century. In 1866 Erichsen attributed the nervous symptoms exhibited by victims of railway accidents to neurological mechanisms, concluding that their traumatic shock was caused by neurological effects resulting from concussion of the spine. Charcot moved on from this belief to label these symptoms 'hysteric', connecting the symptoms seen in 'railway spine' with a type of psychological trauma (Young, 1995, pp. 5, 16, 19). Although other theories appeared in relation to 'railway spine', such as by the German doctor Herman Oppenheim who labelled it a 'traumatic neurosis', the relationship between increasing railway injuries in the nineteenth century and hysteria posited by Charcot, and later supported by Janet and Freud in the 1890s, relied on the key elements of the violent event (the railway accident) and traumatic memory, leading to psychological symptoms. Importantly, Charcot also integrated gender into his reasoning, asserting that women were more vulnerable to the condition than men because of their specific constitution (Young, 1995, p. 20). This point is relevant to the application of these ideas to war-related trauma among predominantly male soldiers, who hold a specific position within the Australian national story that is not synonymous with vulnerability.

While Janet and Freud were developing their own ideas related to the traumatic memory resulting from the violent event and moving decidedly away from physical explanations for 'psychic' events, early Australian psychiatrists, most of whom worked within the asylum system, remained attached to moral therapy. This approach assumed that hard work and productiveness were essential qualities in the civilised (male) citizen and so attempted to restore this key figure through predominantly physical treatments (Coleborne, 2015, p. 118). The incongruity between masculinity and madness, as earlier raised by Charcot, meant that the insane male was treated as an aberration. This intensified in the Federation period, when male citizens, represented publicly through the figure of the rugged, hardworking and independent 'bushman', were positioned as representatives of the strong new Australian nation. When some Australian soldiers, many of whom had lived their lives as bushmen, returned from the South African War of 1899–1902 exhibiting traumatic symptoms, their role as martial—and thus hyper-masculine—representatives of the nation saw attempts by doctors to distance any notion of madness from their condition. Medico-cultural interpretations of their condition saw an insistence on physical causes for psychological disorders, in keeping with prevailing knowledge of the time or outright denial of their illness (Karageorgos, 2020). The British establishment of a hospital specifically for traumatised soldiers from 1869 in Netley where South African War veterans were treated, suggests that Australian approaches were less advanced than those appearing in other parts of the world (Jones & Wessely, 2005, pp. 7–8).

The approach to shell shock emerging from the First World War (WWI) appeared to veer away, in practice at least, from the traumatic memory, influenced by Freud's developing theory of the neurosis, which he attributed not only to the crucial event but also to a person's early childhood experiences (Young, 1995, pp. 37–38). Medical men, seeking to legitimise their authority in the years before WWI, pointed to their 'privileged access' to the traumatic memory in mentally ill individuals. However, even internationally, very few physicians integrated this rationale into their treatment of traumatised WWI soldiers, focusing more on organic causes for mental diseases such as shell shock, neurasthenia, hysteria and disordered action of the heart (DAH) (Young, 1995, pp. 38, 40, 50). Others also pointed to heredity, with the British doctor Frederick Mott, for example, attributing war neurosis to 'an inborn timidity or neuropathic disposition or an inborn germinal or acquired neuropathic or psychopathic taint' (Mott, as cited in Young, 1995, p. 54). Australian physicians working on both the war and home fronts were similarly attached to somatic explanations for psychological symptoms, however, were more likely than European authorities to label those who succumbed to the mental pressures of soldiering 'morally weak', appearing to point more often to heredity as one cause for their weakness or lack of morality (Tyquin, 2006, pp. 34–35; Bourke, 1996, p. 117).

Australian attitudes towards the traumatised soldier had more in common with German approaches that focused more on soldiers' fears in explaining mental symptoms than their wartime experiences (Fassin & Rechtman, 2009, p. 62). Australian military-medical authorities also did not begin invaliding men with shell shock until 1916 despite working alongside the British Army Medical Corps who did so from 1914 (Tyquin, 2006, pp. 24–25). Although some Australian battlefield doctors, including the Victorian psychiatrist John Springthorpe, integrated Freudian techniques into their treatment of WWI soldiers, the vast majority were decidedly attached to morality, heredity and 'social infirmity' (Damousi, 2005, p. 37; Tyquin, 2006, p. 34).

Heredity and Personality in the Second World War

Even in Britain, the move towards psychology and away from moral or personal explanations for war-related trauma was not reflected in the conceptualisation of 'traumatic neurosis' during the Second World War (Bourke, 1995, pp. 115, 118). Australian military authorities followed the example of the United States in adopting psychological principles to screen enlisting recruits (Pols, 2007). Psychological testing across the military branches began to be discussed in May 1941, when an Advisory Committee on Psychology Testing was instituted, leading to the appointment of Major Hugh Fowler as Deputy Assistant Adjutant-General in the Directorate of Recruitment and Mobilization and commencement of formal testing of recruits in September 1942. Although this broader approach came relatively late in the war, due to the 'lack of general knowledge and acceptance of Vocational and Industrial Psychology by the community at large', psychological screening occurred at the state level in Victoria, Western Australia, Queensland and New South Wales as early as November 1940 (Australian Army Psychology Service, 1940). Although psychological testing was acknowledged as useful to test intelligence and personality and predict the likelihood of any possible mental breakdown, the Director-General of Medical Services Major-General Rupert Downes wrote on 31 August 1940 that its 'sphere of usefulness in the services' was limited, perhaps strategically pointing to the superior knowledge among his own Australian Army Medical Service psychiatrists in the 'elimination of personnel who are mentally unfit on enlistment' (Downes, 1940). In Australia, as elsewhere, the military emphasis on productivity and personality testing expanded to include civilian populations when the post-war economic boom increased the demand for 'productive workers' (Cohen, 2016, p. 101).

Despite Downes' words, military psychiatrists were not automatically integrated into military units from the early years of the war, despite the continuing need to support traumatised WWI soldiers in the interwar period. Peter Hobbins (2007, p. 31) points to the conspicuous absence of sufficient psychiatric nomenclature in military archives from the early years of the Second World War, reflecting the opinion within military circles that implanting ideas about trauma in the minds of fearful soldiers could make them more likely to succumb to its symptoms. Reports written by military-medical authorities throughout the war positioned neurotic symptoms and disorders primarily within a moral and social framework. The terms of this framework were set out on 29 May 1940, when the Director-General of Medical Services, Major-General Downes, convened a conference with representatives from the Army, Repatriation Commission and the Royal Australasian College of Physicians to discuss the best way to address psychiatric symptoms or conditions among soldiers (Walker, 1952, p. 676). In this important forum, Downes (1940) emphasised that 'the war neuroses are to be regarded as manifestations of weakness and failure of morale rather than inevitable battle casualties'. The conference attendees did acknowledge that 'there will be more cases probably in this war than in the last war', but went on to intentionally establish a distance between the clinical and moral. They did so by noting that military-medical staff 'must be severe in decision on what cases are really neuroses' and would be required to 'do all possible to remove the attitude of sentimentality to the soldier', thus displaying an awareness of their own influence on the diagnosis of mental trauma beyond the clinical, and a desire for its moral understanding to influence the way it was experienced by soldiers. Therefore, in attempting to ensure that cases of war neurosis do not overwhelm the military and allow them to retain manpower, there was a clear attempt made to initially de-medicalise the traumatised soldier.

Although psychiatric care away from the battlefield was recognised by military authorities as necessary for more severe war neurosis cases, there was an intentional sidelining of psychiatric knowledge and treatment on the battlefield and in surrounding areas. Traumatised men were initially cared for by Field Ambulance units, who were instructed to keep men as close to the battlefield as possible, based on a belief that the longer a man was away from his fighting unit, the more likely that his neurosis would worsen, making him unable to return to duty. If the situation was not resolved quickly, they would be moved to a Casualty Clearing Station, and then if the patient was not improved within three days, an Australian General Hospital (AGH) (Walker, 1952, p. 684; Hobbins, 2007, p. 31).

In their report on the 9th Division AIF in Tobruk, military psychiatrists E.L. Cooper and A.J.M. Sinclair (1941) specified that while mental cases remain in the AGH, medical examinations should 'establish in the mind of the patient an idea that he is suffering from obscure organic disease', reflecting the fact that many cases of perceived neurosis on the war front did have somatic causes. Even at this stage, the need for psychiatric care is not mentioned. Rather, 'occupational therapy, re-education, explanation [which suggests psychotherapeutic support] and other measures' were emphasised while 'the man can be trained once more to become an efficient soldier' (Cooper & Sinclair, 1941).

If soldiers were proven to be 'war neurotics', meaning that the attempts at de-medicalisation of their condition as described above failed to produce results, this meant that they were brought before a Medical Board for assessment and likely repatriated to Australia. At this stage the influence of 'personal and family history' was prioritised in assessing the patient; however, this was a consideration within the military from the enlistment period. For example, Australian Lieutenant-Colonel Macintosh wrote in a November 1940 instructional document on recruit examinations that medical examiners needed to think along military rather than civil lines when assessing recruits, emphasising the danger of 'false pension claims' and urging the examiner to remember that 'the army has great attraction for the undesirable types, especially the subnormal and even mental types' (Macintosh, 1940). Cooper and Sinclair (1941) wrote that if soldiers were accepted into the military forces and developed neuroses, 'the "previous personality" of the patient prior to the present stress was of considerable importance in determining the onset, the texture and severity of the psychoneurosis'.

There is no doubt throughout the report by Cooper and Sinclair (1941) where the priorities of the military lie, stating that 'the primary function of the medical services is the maintenance of the strength of the fighting forces'. They specified that that the soldier should be kept in the divisional area if he is 'capable of rendering further military service without adversely influencing his fellow soldiers', writing that the aim of these efforts is ultimately the 'rehabilitation of the individual as a soldier and as a man'. The importance of rehabilitating the neurotic as both 'a soldier and a man' suggests that a soldier can only be effective if they demonstrate features of effective masculinity. The gendering of the rehabilitated soldier is important in the context of Australian military psychiatry, which during the Second World War focused more on the social positioning of the male neurotic soldier than United States military-medical authorities, who were more likely to attribute their condition to 'exhaustion and fatigue' (Garton, 1996, p. 166). Interestingly, in 1949 Allan Walker sent a draft of his official medical history of the conflict to the journalist John F. Williams for comment, who wrote that 'some of our anti-feminists' may be 'rather doubtful as to the truth of the paragraph regarding the feminine nervous system being less vulnerable' to psychiatric disorders (Williams, 1949). Gendered visions of mental illness were common across the civilian and military worlds, but were a key consideration when applied to the soldier, who was expected to represent the hyper-masculine ideal (Garton, 1988, p. 114).

As late as August 1942, Royal Australian Air Force (RAAF) authorities felt that local medical officers integrated into the military were more efficient than a 'metropolitan psychiatrist to enquire into and assess the mental stresses to which their own personnel are begin subjected', referring again to the perceived moral nature of trauma symptoms by saying that this applied 'particularly to pilots whose moral fibre is under suspicion' (RAAF, 1942). The Regimental Medical Officer (RMO) was certainly key to both prevention and treatment of men displaying symptoms of war neurosis. However, the emphasis on the RMO as the best person to address cases of trauma in a way that ensured minimal loss of manpower reflects the continuing emphasis not only in Australian military circles but also among allied British and American forces, on the prevention of malingering as a key priority for all treating soldiers with traumatic symptoms (Cooper & Sinclair, 1941; Hobbins, 2007, p. 31; Fassin & Rechtman, 2009, p. 68; Karageorgos, 2023).

Although weakness, social incompetence or abnormal family history was attributed to the traumatised soldier during the First World War, these ideas appeared to gain more prominence in the minds of Australian military doctors during the Second World War and were key to Allan Seymour Walker's official Australian medical history of the conflict, first published in 1952 (Garton, 1996, pp. 163, 166). In this volume, Walker acknowledges that approaches towards war neurosis from the First World

War were inadequate and soldier treatment was influenced by the continuing association of psychiatry with the 'madhouse', but he is reluctant to attribute military trauma to the violent events of war, writing that 'unhappily today' some psychiatric approaches include a 'suggestion, even a belief, that damage through the violence of war is alone responsible for the mental disorders of servicemen and servicewomen' (Walker, 1952, p. 675). In attempting to distance traumatic disorder from war itself, Walker's history points to a variety of factors including heredity, delinquency, social inadequacy and even geography as more influential on the mental state of a soldier (Walker, 1952, pp. 678, 679).

Such an emphasis on faulty personality or heredity continued into the 1960s and 1970s. My own historical work which examines the negative comments made about traumatised Second World War and Korean War veterans by a group of repatriation doctors during this period shows that legitimacy is only accorded to the traumatised soldier if their 'deviance' as a sick person aligns with medical knowledge (Karageorgos, 2023). References to the 'phony crony' by these doctors reflected their role as detectives of malingering and arbiters of repatriation policy. From 1962 until 1972 Australian soldiers were engaged in the Vietnam War, a conflict that introduced new form of war neurosis which was officially labelled as post-traumatic stress disorder (PTSD) in the DSM-III of 1980. The categorisation of PTSD is significant within a general narrative of military-medical responses to war trauma, as it marks a point at which the clinical approach to war trauma officially moves away from heredity and back towards the traumatic memory related to the violent event.

DSM-III and the Traumatic Event

The DSM-III of 1980 was the first that adopted a neo-Kraepelinian approach to the diagnosis and categorisation of psychiatric categories. The use of a classification system for mental illness, popularised by the German psychiatrist Emil Kraepelin in the late nineteenth century, was decided upon by the architects of DSM-III led by Robert Spitzer, meaning that psychiatric conditions were to be categorised by symptoms serving as diagnostic criteria. Although this approach fell out of favour in the mid-twentieth century as psychological explanations for mental disorder overtook biological aetiology, the introduction of psychoactive drugs from the 1950s made a case for the return to biology in the production of mental health categories (Young, 1995, pp. 95–97). DSM-III was also the first psychiatric handbook shaped by specifically 'neo-liberal obsessions with efficiency, productivity and consumption', meaning that it reflected prominent ideological norms of the period and aimed at producing citizens that could align with those norms (Cohen, 2016, p. 79).

The inclusion of PTSD in DSM-III, which officially shifted the popular approach to diagnosing and treating war-related trauma, is an important case study illustrating the politicisation of mental health categorisation. As sociologist Wilbur Scott (1990) writes: 'the struggle for *recognition* of PTSD by its champions was profoundly political' (p. 295; original emphasis). This began with a 1970 visit by anti-war psychiatrists Chaim Shatan and Robert Lifton to the New York City chapter of Vietnam Veterans Against the War (VVAW). Their inclusion in the chapter's 'rap groups' which, in Lifton's words, facilitated 'street-corner psychiatry' did not only have a therapeutic aim but also a political one—recording Vietnam experiences to share with the seemingly oblivious American public (Scott, 1990, p. 300). In 1971, Shatan published an op-ed in the *New York Times* about 'post-Vietnam syndrome', which he formulated based on his recognition of a 'delayed massive trauma' in VVAW rap groups with symptoms such as 'guilt, rage, psychic numbing, alienation' and 'feelings of being scapegoated' (Scott, 1990, p. 301; Young, 1995, p. 109). In 1973 Lifton published *Home from the War*, based on his own observations with VVAW and resultant criticism of established military psychiatry methods. The joint association of Shatan and Lifton with the National Council of Churches (NCC) led to their involvement in the NCC-planned First National Conference on the Emotional Needs of Vietnam-Era Veterans, in which 130 veterans, psychiatrists, chaplains and Veteran's Administration (VA) representatives met in St. Louis to discuss readjustment issues in Vietnam veterans. Also in attendance was Sarah Haley, a Boston VA social worker whose work with veterans who had been involved in the massacre of Vietnamese civilians at My Lai in 1968 had caused her to identify the violent events of war as the source of their trauma (Young, 1995, p. 109). Although her VA colleagues overtly shunned this idea, she continued to advocate for veterans, including providing VVAW chapters with lists of recommended VA branches that would be more sympathetic to their condition (Scott, 1990, p. 299). Her association with Lifton through a shared concern about the impact of My Lai on veterans led to their joint membership of the Vietnam Veterans Working Group (VVWG), along with Shatam as well as other veterans and clinicians.

When, in 1975, Spitzer began collating what would become DSM-III, Lifton and Shatam requested that he form a sub-committee to specifically investigate 'post-Vietnam syndrome'. This was initially rejected as Spitzer had received advice that psychiatric disorders in Vietnam veterans did not need a specific diagnostic classification. The biological psychiatric approach favoured by Spitzer remained attached to heredity and personality as explanations for traumatic symptoms in soldiers and veterans. In fact, he was assured by Lee Robbins from Washington University that Vietnam veterans were all 'character disorders', who 'came from rotten backgrounds', stating that they 'were going to be malcontents and dysfunctional anyway … Vietnam is not the cause of their problems' (Robbins, as cited in Scull, 2019, p. 222). However, Lifton and Shatam, with other members of the VVWG, realised that they needed to encourage societal interest in the issue, and did so by publicising the existence of post-Vietnam syndrome in the media, in academic journals, within the American Psychiatric Association (APA) and the American Orthopsychiatric Association (AOA) (Scott, 1990, p. 305). At this stage, according to Young, mental illness among Vietnam veterans became a 'politically sensitive issue', meaning that Spitzer was persuaded to form a working group, leading eventually to the inclusion of post-Vietnam syndrome, by then renamed to PTSD as its symptoms were recognised as not specific to veterans, also affecting people involved in natural or man-made catastrophes (Young, 1995, pp. 110, 111; Horwitz, 2021, p. 81). This was the first available nosology of a psychiatric condition affecting veterans, which also, at least officially, spelled the end of heredity and faulty personality within the aetiology of war-related trauma. But it was also one that would not have existed without a moral and political incentive to recognition.

The DSM-III definition of PTSD bases its diagnostic criteria on the 'psychologically traumatic event that is generally outside the range of usual human experience' and the resulting symptoms, which include 'reexperiencing the traumatic event; numbing of responsiveness to, or reduced involvement with, the external world; and a variety of autonomic, dysphoric, or cognitive symptoms' (APA, 1980, p. 236). DSM-I of 1952 included a precursor of sorts to PTSD, notably 'gross stress reaction', defined as a temporary condition possibly caused by 'combat' or 'civilian catastrophe' that was caused by 'great or unusual stress' in which a 'normal personality may utilize established patterns of reaction to deal with overwhelming fear' (APA, 1952, p. 40). In DSM-II, published in 1968, the condition 'transient situational disturbance' was included, which was described as an 'acute reaction to overwhelming environmental stress' and which could include 'fear associated with military combat and manifested by trembling, running and hiding' (APA, 1968, p. 49). It is notable that neither of these definitions acknowledged that traumatic symptoms from war could become chronic or delayed, focusing solely on their transitory and immediate nature (Horwitz, 2021, p. 80).

The focus on the traumatic memory based on a violent event in DSM-III, then, was a partial return to the nineteenth-century ideas of Charcot, Freud and Janet, among others, and specifically Freud's 1919 concept of the 'war neuroses' which he claimed occurred 'after frightening experiences or severe accidents'. He writes in the introduction to *Psychology and the War Neuroses* authored by Freud along with psychoanalysts Sándor Ferenczi, Karl Abraham, Ernst Simmel and Ernest Jones that the war neuroses 'may proclaim too loudly the effects of mortal danger' from which the 'human ego is defending itself' (Freud, 1955, pp. 208, 209). The traumatic memory itself is a 'manmade object' and its transformation since the nineteenth century is dependent on the 'enlargement of the memory's active principle, the pathogenic secret' into the collectivity. The expansion of the 'pathogenic secret' means that it needed to be collectively recognised, which occurred via a political act in the post-Vietnam War period when veterans and their supporters demanded attention (Young, 1995, p. 142). There is significant evidence in the archives that this demand expanded to include Australian veterans in the post-DSM-III period; however, this branch of veteran activism lies outside the scope of this chapter.

Significantly, the DSM-III definition of PTSD specified that it was based on the 'existence of a recognizable stressor that would evoke significant symptoms or distress in almost everyone', meaning that there was no reference to family or personal history in determining the existence of the condition (APA, 1980, p. 238). From DSM-IV, published in 1994, the phrases 'almost everyone' and 'traumatic event … outside the range of normal human experience' were omitted, and instead, it said that 'the person experienced, witnessed or was confronted with events that involved actual or threatened death or serious injury, or a threat to the physical integrity of self or others' and 'the person's response involved intense fear, helplessness or horror', thereby increasing the reach of PTSD to include people who did not personally experience the traumatic event (APA, 1994, pp. 431–432, 427–428). In DSM-5, the diagnostic criteria and accompanying explanation is significantly more detailed than in any previous DSM; however, the listed criteria have served to decrease the range of possible traumas and specify that trauma can only occur in response to a definite event (APA, 2013, pp. 271–272; Horwitz, 2021, pp. 154–155).

Although the approach to PTSD in the DSM since 1980 has been imperfect and reduces a broad spectrum of symptoms to a limited range of categories, its inclusion in a text that has 'come to completely dominate our society's approach to mental illness' is significant (Scull, 2019, p. 274). In his recent history of the DSM, Horwitz (2021, p. 149) writes that the inclusion of diagnostic categories 'made mental disorders seem more *real* to the public, to physicians in other medical specialties, to insurance companies, and to federal regulators'. This includes the DVA, which continues to align its policies with the DSM. Despite the reversal of support by some psychiatric figures and organisations internationally upon the publication of DSM-5 in 2013, based on some cases on an objection to the reliance on a purely medical, or biological, approach to mental illness, the DVA Guidelines for Psychiatric Compensation Claims specifies that 'before a finding that a claimant has a particular psychiatric condition, the claims assessor must ensure that all of the diagnostic criteria in DSM-5 are addressed'. Advice provided by the DVA regarding the structure of the psychiatric report does specifically focus on 'stressful events' that were 'experienced, witnessed or learned about by the claimant, before, during or after military service that could be aetiologically relevant for any diagnosed psychiatric disorders'. Significantly, it also asks for information about the veteran's 'history of substance abuse' and 'family psychiatric history' (DVA, 2019), which can be included as pretraumatic factors influencing the development of PTSD according to DSM-5 (APA, 2013, p. 277).

The Royal Commission into Defence and Veteran Suicide

Since the late twentieth century, the ADF and Australian Government have taken significant steps to investigate the seemingly worsening mental health crisis in the military. This has included the April 2019 *Inquiry into Transition from the Australian Defence Force (ADF)*, which found that the 'mental health stigma is very much alive and well' and the fear of this stigma means that individuals do not report psychological issues or access mental health care for fear of disappointing colleagues or risking their position in the military hierarchy (Joint Standing Committee on Foreign Affairs, Defence and Trade, 2019, p. 42; see also AIHW, 2017). Importantly, the Inquiry also found that although

> the upper levels of rank and command get it – they understand ... there are people in the middle section who don't get it. Maybe they have their own prejudice against mental health or whatever and so it will depend on who that person's direct line manager is. That makes a big different to how they're going to be looked after. (Romaniuk, as cited in Joint Standing Committee on Foreign Affairs, Defence and Trade, 2019, p. 43)

The attention paid here to both the attitudes of colleagues and mid-level officers suggests that attitudes about mental health in the ADF are not only set by the organisation itself but also reliant on a person's general attitudes to military-psychological issues, which arise from their experiences both within the forces and in society as a whole. Thus, addressing stigma against mental health issues within the military requires both the ADF and political authorities to move beyond simply military training, to a whole-society approach.

Bound up in these ideas about the psychology of military personnel—both within and beyond the ADF—are gendered ideas about the male soldier and the type of masculinity they are expected to embody. David Duriesmith's recent work on the Australian Special Forces in Afghanistan highlights the social basis of violent military masculinities, which he believes 'exists in relation to and in continuity with mainstream masculinities that centre violence, toughness, aggression and fraternal bonding'. Duriesmith (2023, pp. 282–283) insists that addressing these masculinities requires militaries to look beyond individual military units to masculinities both within the ADF and Australian society.

The *Inquiry into Transition from the Australian Defence Force* (ADF) has taken place among a series of other interventions by the ADF that seek to reduce mental health issues among serving soldiers and veterans, which includes the aforementioned *Defence Mental Health and Wellbeing Strategy 2018–2023*. From a clinical perspective, given the ADF's alignment with the DSM as mentioned above and the continual attention to military mental health, it would be expected that any ideas about heredity or personality would be minimised. However, the narrative emerging thus far within the publicly available submissions to the Royal Commission into Defence and Veteran Suicide suggests that long-standing societal and military stigmatisation of military trauma persist. The remainder of this chapter will outline a collection of these submissions and their alignment with historical understandings of trauma.[1]

Erving Goffman traces the source of the word 'stigma' to the Greeks, who saw it as 'bodily signs designed to expose something unusual and bad about the moral status of the signifier'. He goes on to identify a process within the life of a stigmatised person, in which they 'incorporate the stand-point of the normal' or the 'identity beliefs of the wider society' and thereby become aware of the status conferred by possessing a particular stigma (Goffman, 1968, pp. 11, 45). Respondent 1 (2021)

[1] Responses are anonymised to protect individual persons; however, it is noted that each publicly available submission has been cleared by its author for public distribution.

writes in his submission to the Royal Commission of the 'stigma of members seeking help within the "rehabilitation platoon"'. This included the 'disgusting' treatment of members who had attempted suicide by the 'chain of command and senior members', by which they were 'considered weak and punished ostracised and laughed at', writing that 'we as junior members had to take it on ourselves to ensure these members were ok'. He also writes that members in the rehabilitation platoon were forced to do guard duty three to four times per week and were told they had to do it because they were 'lingers'. This reference to malingering is important, as it has been a concern within the Australian military since its first nation-wide campaign in South Africa in Federation period. Moreover, accusations of malingering have frequently been directed towards traumatised soldiers; however, battlefield doctors who were directed to identify malingerers during the world wars frequently chose to label malingering as a medical condition caused by heredity or psychology rather than a criminal offence, so as to 'keep their medical hat on' (Kanaan & Wessely, 2010, pp. 75–76; Karageorgos, 2023). The DSM-5 defines malingering as the 'intentional production of false or grossly exaggerated physical or psychological symptoms', which includes 'avoiding military duty' (APA, 2013, p. 726). The submission by Respondent 1 when viewed alongside the historical pathologisation of malingering reveals the persistence of views that separate malingering and trauma, and ultimately—as identified by the respondent—perpetuate stigma within the ADF.

Other submissions to the Royal Commission connect the awareness by soldiers of the stigma against mental health issues with the fear of risking their careers in reporting. Respondent 2 (2021) discusses over-using alcohol during a 17-month investigation by the ADF into an alcohol-related incident that he did not commit, which caused him to become depressed. The respondent writes that his workload was doubled or tripled by his unit in an attempt to 'break' him, which led to deep feelings of shame 'not only as a soldier but as a father and husband'. He stated that the decision to discuss his relationship with alcohol with his Company Sergeant Major was used as a reason to terminate him, with his career manager writing that he 'chose to consume alcohol to deal with the stress of an investigation'. This has led to three suicide attempts. Notable here is the decision to write that the soldier 'chose' to turn to alcohol. I have earlier written about the tendency by Repatriation Department doctors in the 1960s and 1970s to de-medicalise traumatised veterans who drink by attaching moral judgements onto them (Karageorgos, 2023). However, alcoholism is a common symptom of war-related trauma (Garton, 1996, p. 95). After discussing harrowing war front experiences in East Timor and Iraq, Respondent 3 (2021) writes that despite both physical and psychological issues resulting from his service, including PTSD and anxiety, he was reluctant to report because the 'culture in the Army at the time was to try your hardest not to medically discharge as it was thought to be weak, and I did not want it on my record because I thought it may jeopardise my future employment', meaning that he stayed silent about any issues he was experiencing.

Respondents to the Royal Commission frequently discuss the role of masculine norms in determining both their behaviour and their treatment within the military. After highlighting the increased risk of veteran suicide among men, Respondent 4 (2021) writes that there are 'some distinctly "man" conditions and behaviours that are obviously massive risk factors here'. A range of submissions discuss comparisons made between combat and non-combat soldiers, and suggest that soldiers who do not experience combat are considered immune to PTSD due to the physical and mental difficulty of combat. For example, Respondent 5 (2021) was told by a DVA representative that he was 'ridiculous' and that it was 'impossible that someone from a non-combat corp … could claim to have PTSD … and I needed to be in infantry or something', which led to suicidal ideation. After being diagnosed with PTSD and Major Depressive Disorder following a suicide attempt, Respondent 6 (2021) was similarly told by his DVA Case Manager that 'only soldiers get PTSD'. However, the definition of PTSD in DSM-5 does not require a person to have personally observed a traumatic event to develop the condition but can occur as a result of 'learning that the traumatic event(s) occurred to a close family member or close friend' or 'experiencing repeated and extreme exposure to aversive details of the traumatic event' (APA, 2013, p. 271). It is significant that these ideas came from the DVA, the ultimate political arbiter in the distribution of scarce public funds as pensions. The 'moral genealogy' of war-related trauma is not restricted to the military setting but draws knowledge from across society and 'can be read in various ways, depending on the political purpose it serves' (Fassin & Rechtman, 2009, p. 209). Respondent 6 was evidently aware of this, ending his submission with the hope that 'the findings of this Royal Commission will initiate positive change', although he was 'apprehensive that the current [Liberal Party] leadership is the one to do it'.

Importantly, respondents to the Royal Commission frequently commented on the disregard or 'negligence' displayed by military superiors when they were told of mental health issues, which caused delays that constituted 'failings of duty of care'. Respondent 7 (2021) writes that once he had 'mustered up the courage to reach out and ask for help', he was faced with a lack of concern from the corresponding unit responsible for mental health care, Navy Psychology West:

> [The Department of] Defence is attempting to breakdown [sic] the stigma surrounding mental health and advocate multiple support mechanisms...yet when I attempted to use one of the support services I was told 'no, wait for your annual psych appointment' and had the door closed on me because I wasn't 'high risk'.

Such delays can be life-threatening to soldiers or veterans in an urgent psychological state.

Conclusion

Viewing representative submissions to the Royal Commission into Defence and Veteran Suicide alongside a historical-psychological analysis reveals that the 'clinical genealogy' of war-related trauma, which includes formal procedures followed by DVA-appointed psychiatrists, does not alone determine the treatment of some soldiers and veterans by the DVA or individuals within the ADF. As mentioned, the *Inquiry into Transition from the Australian Defence Force (ADF)* shows that reluctance exists within the 'middle section' of the military hierarchy to respond supportively to psychological issues, and suggests that ideas about heredity and personality rooted in twentieth-century perceptions of men and combat persist. The submissions to the Royal Commission cited above suggest that the 'moral genealogy' of trauma, developed beyond the clinical setting, still influences conceptions of trauma within both the ADF and DVA, illustrating that 'trauma has far exceeded the grasp of psychiatrists and their debates about how to define it' (Fassin & Rechtman, 2009, p. 153). Recognising that such ideas persist and continue to be applied to the traumatised soldier or veteran by political and military authorities provides a basis for continued historical engagement with the findings of the Royal Commission.

This chapter has outlined clinical approaches to military and other types of trauma since the late nineteenth century. Although some clinicians at the turn of the twentieth century saw mental symptoms as emerging from external stimuli, namely, the traumatic memory of a violent event, the view of traumatic response to war as a hereditary trait or fault in personality became prominent until it peaked in the period surrounding the Second World War, when despite overwhelming evidence of a connection between combat and mental illness, military-medical staff were instructed to initially de-medicalise traumatised soldiers to prevent a loss of manpower. The publication of DSM-III in 1980 saw a return to the external and violent event, and focused on establishing diagnostic criteria to categorise a new condition, PTSD.

Despite understandings of war-related trauma promulgated in the DSM and its versions since 1980, outdated views of military mental health exist both within and outside of the ADF. In addressing problems within any 'regime of truth', Foucault (1977, p. 14) offers a solution: 'detaching the power of truth from the forms of hegemony (social, economic, and cultural) within which it operates at the present time'. Moving past established 'truths' about traumatised soldiers within ADF culture will require a commitment to addressing society-wide perceptions of the soldier and veteran across Australian society.

References

Alexander, J. C. (2012). *Trauma: A social theory*. Polity.

American Psychiatric Association. (1952). *Diagnostic and statistical manual of mental disorders*. APA.

American Psychiatric Association. (1968). *Diagnostic and statistical manual of mental disorders* (2nd ed.). APA.

American Psychiatric Association. (1980). *Diagnostic and statistical manual of mental disorders* (3rd ed.). APA.

American Psychiatric Association. (1994). *Diagnostic and statistical manual of mental disorders* (4th ed.). APA.

American Psychiatric Association. (2013). *Diagnostic and statistical manual of mental disorders* (5th ed.). APA.

Anderson, B. (1983). *Imagined communities: Reflections on the original and spread of nationalism*. Verso.

Australian Army Psychology Service. (1940). *Description of the Australian Army Psychology Service*. Psychological Methods and Efficiencies of the Fighting Forces, AWM54 804/1/2, Australian War Memorial, Canberra.

Australian Institute of Health and Welfare. (2017). *Incidence of suicide among serving and ex-serving Australian Defence Force personnel 2001–2015: In brief summary report*. AIHW.

Bourke, J. (1996). *Dismembering the male: Men's bodies, Britain and the Great War*. Reaktion Books.

Braslow, J. (1997). *Mental ills and bodily cures: Psychiatric treatment in the first half of the twentieth century*. University of California Press.

Cohen, B. M. Z. (2016). *Psychiatric hegemony: A Marxist theory of mental illness*. Palgrave Macmillan.

Coleborne, C. (2015). *Insanity, identity and empire: Immigrants and institutional confinement in Australia and New Zealand, 1873–1910*. Manchester University Press.

Cooper, E. L., & Sinclair, A. J. M. (1941). *War Neuroses in Tobruk: A Report on 207 Patients from the Australian Imperial Force Units in Tobruk*. AWM54 267/6/30. Australian War Memorial, Canberra.

Damousi, J. (2005). *Freud in the Antipodes: A cultural history of psychoanalysis in Australia*. UNSW Press.

Department of Defence. (2017). *Defence mental health and wellbeing strategy 2018–2023*. Commonwealth of Australia.

Department of Veterans' Affairs. (2019). *CM7014 MRCC181 – Guidelines for psychiatric compensation claims*. DVA. https://clik.dva.gov.au/compensation-and-support-reference-library/commission-guidelines/cm7014-mrcc181-guidelines-psychiatric-compensation-claims

Dixson, M. (1999). *The imaginary Australian: Anglo-Celts and identity 1788 to the present*. UNSW Press.

Downes, R. (1940). *Psychological methods and efficiencies of the fighting forces*. AWM54 804/1/2, Australian War Memorial, Canberra.

Duriesmith, D. (2023). Transforming masculinities after scandal: The response to Australia's war crimes in Afghanistan and the possibility of change in military masculinities. *Australian Journal of Political Science, 58*(3), 272–286. https://doi.org/10.1080/10361146.2023.2223492

Fassin, D., & Rechtman, R. (2009). *The empire of trauma: An inquiry into the condition of victimhood*. Princeton University Press.

Foucault, M. (1977). The political function of the intellectual. *Radical Philosophy, 17*, 12–14.

Freud, S. (1955). Introduction to psycho-analysis and the war neuroses. In *The standard edition of the complete psychological works of Sigmund Freud*. Hogarth Press.

Garton, S. (1988). *Medicine and madness: A social history of insanity in New South Wales, 1890–1930*. UNSW Press.

Garton, S. (1996). *The cost of war: Australians return*. Oxford University Press.

Goffman, E. (1968). *Stigma*. Penguin.

Hobbins, P. G. (2007). 'Living in hell but still smiling': Australian psychiatric casualties of war during the Malaya-Singapore campaign, 1941–42. *Health and History, 9*(1), 28–55. https://doi.org/10.1353/hah.2007.0003

Holbrook, C. (2014). *Anzac: The unauthorized biography*. NewSouth.

Horwitz, A. V. (2021). *DSM: A history of psychiatry's bible*. John Hopkins University Press.

Joint Standing Committee on Foreign Affairs, Defence and Trade. (2019). *Inquiry into transition from the Australian Defence Force (ADF)*. Joint Standing Committee on Foreign Affairs, Defence and Trade.

Jones, E., & Wessely, S. (2005). *Shell shock to PTSD: Military psychiatry from 1900 to the Gulf War*. Psychology Press.

Kanaan, R. A. A., & Wessely, S. C. (2010). The origins of factitious disorder. *History of the Human Sciences, 23*(2), 68–85. https://doi.org/10.1177/0952695109357128

Karageorgos, E. (2020). The bushman at war: Gendered medical responses to combat breakdown in South Africa, 1899–1902. *Journal of Australian Studies, 44*(1), 18–32. https://doi.org/10.1080/14443058.2020.1717581

Karageorgos, E. (2023). Medical fears of the malingering soldier: "Phony cronies" and the Repat in 1960s Australia. *Medical History, 67*(2), 172–191. https://doi.org/10.1017/mdh.2023.19

Lake, M., & Reynolds, H. (2010). *What's wrong with Anzac? The militarisation of Australian history*. NewSouth.
Lifton, R. J. (1973). *Home from the war: Vietnam veterans: Neither victims nor executioners*. Simon and Schuster.
Macintosh. (1940). *Medical examinations (recruits)* (AWM54 317/3/1). Australian War Memorial, Canberra.
Pols, H. (2007). War neurosis, adjustment problems in veterans, and an ill nation: The disciplinary project of American psychiatry during and after World War II. *Osiris, 22*, 72–92. https://doi.org/10.1086/521743
Respondent 1. (2021, December 9). *Submissions to the royal commission into defence and veteran suicide*. Response ID ANON-Z1e7-Q8HY-H. https://defenceveteransuicide.royalcommission.gov.au/
Respondent 2. (2021, September 29). *Submissions to the royal commission into defence and veteran suicide*. Response ID ANON-Z1e7-Q15U-K. https://defenceveteransuicide.royalcommission.gov.au/
Respondent 3. (2021, December 27). *Submissions to the royal commission into defence and veteran suicide*. Response ID ANON-Z1E7-Q8DS-7. https://defenceveteransuicide.royalcommission.gov.au/.
Respondent 4. (2021, December 21). *Submissions to the royal commission into defence and veteran suicide*. Response ID ANON-Z1E7-Q81T-N. https://defenceveteransuicide.royalcommission.gov.au/
Respondent 5. (2021, November 30). *Submissions to the royal commission into defence and veteran suicide*. Response ID ANON-Z1E7-Q8P9-S. https://defenceveteransuicide.royalcommission.gov.au/
Respondent 6. (2021, July 16). *Submissions to the royal commission into defence and veteran suicide*. Response ID ANON-Z1E7-Q1GT-4. https://defencev-eteransuicide.royalcommission.gov.au/
Respondent 7. (2021, September 22). *Submissions to the royal commission into defence and veteran suicide*. Response ID ANON-Z1E7-Q1ZM-G. https://defenceveteransuicide.royalcommission.gov.au/
Rose, N. (2019). *Our psychiatric future: The politics of mental health*. Polity.
Royal Australian Air Force. (1942). Monthly Report of Work of Medical Branch Eastern Area, 11 August. A705 132/1/724 PART 1. Australian War Memorial, Canberra.
Royal Commission into Defence and Veteran Suicide. (2021). *Terms of reference*. https://defenceveteransuicide.royalcommission.gov.au/about/terms-reference
Royal Commission into Defence and Veteran Suicide. (2022). *Interim report*. Commonwealth of Australia.

Scott, W. J. (1990). PTSD in DSM-III: A case in the politics of diagnosis and disease. *Social Problems, 37*(3), 294–310. https://doi.org/10.1525/sp.1990.37.3.03a00020

Scull, A. (2019). *Psychiatry and its discontents*. University of California Press.

Thomson, A. (2013). *Anzac memories: Living with the legend*. Monash University Publishing.

Twomey, C. (2013). Trauma and the reinvigoration of Anzac: An argument. *History Australia, 10*(3), 85–108. https://doi.org/10.1080/14490854.2013.11668482

Tyquin, M. (2006). *Madness and the military: Australia's experience of the Great War*. Australian Military History Publications.

Walker, A. S. (1952). *Clinical problems of war*. Australian War Memorial.

Williams, J. F. (1949). *Letter to Allan Walker*. Psychiatry in the Australian Army – Case Histories and Reports, 1939–1945, AWM54 804/1/4 PART 1, Australian War Memorial, Canberra.

Young, A. (1995). *The harmony of illusions: Inventing post-traumatic stress disorder*. Princeton University Press.

3

Psy-knowledges in the Academy: Making Space for Critique

Emma Tseris

Introduction

In Western contexts, it is widely accepted that we are living in the midst of a 'mental health crisis'. Young people in particular have been identified as experiencing worsening mental health outcomes compared to previous generations (Gao et al., 2023). Claims about rising rates of 'poor mental health', and significant increases in mental distress since the turn of the last century, can be found within contemporary Australian research (Enticott et al., 2022). Underpinning the argument regarding an increasing 'burden' of mental health concerns is the idea that the resources and services within the existing mental health system must be increased (Wykes et al., 2023). This is positioned as an obvious and necessary step to address the needs of an ever-increasing proportion of people who have received a psychiatric diagnosis, even though the existing, growing

E. Tseris (✉)
The University of Sydney, Sydney, NSW, Australia
e-mail: emma.tseris@sydney.edu.au

N. A. Hendry, E. Karageorgos (eds.), *Critical Mental Health in Australia and Aotearoa New Zealand*, The Politics of Mental Health and Illness,
https://doi.org/10.1007/978-3-031-82861-4_3

system may not offer an effective approach to reducing overall rates of distress or mental 'ill-health' (Davies, 2021).

Rose (2008) has charted the rise of psychological discourses as they have achieved the status of 'common-sense' knowledges in many countries across the globe in the twentieth century and the rise of psy-professionals (across a growing array of fields, not only medicine but also in disciplines spanning nursing, education, allied health, and so on). Drawing on biomedical and positivist paradigms of knowledge production, 'mental health' has become a taken-for-granted concept within Australian policy contexts; it is considered to be an objective entity, which can be measured and applied universally (Breen & Darlaston-Jones, 2008). Further, the attainment of mental health or wellbeing is regarded to be an important and responsible component of 'healthy' human development. For example, information produced by the Australian government (Department of Health and Aged Care, 2022, n.p.) outlines:

Good mental health helps you

- cope with the normal stresses of life;
- be productive both at work and in your private life;
- relate well to other people;
- contribute to your community.

Such conceptualisations create tensions and challenges for researching and teaching 'mental health' from a critical perspective. It is challenging to raise critical questions regarding the concept of 'mental health', when it is embedded into curriculum, grant invitations or conference sessions as a taken-for-granted concept. Nevertheless, this is a vital task, given the values that underpin how mental health is conceived within contemporary times. There is a need to radically re-consider the concept of 'mental health' and its limitations as a paradigm for understanding human experience across diverse contexts. For example, it has been noted that the aspiration of 'good mental health' is deeply interconnected with neoliberal citizenship (Cohen, 2016; Esposito & Perez, 2014)—ideas that are evident in the above excerpt, in the image of a 'mentally well' person who has resilience, a willingness to work, and who exerts self-control in their relationships and everyday life. There is a notable silence regarding circumstances including violence, social exclusion, climate disasters, and material deprivation, in which insisting on an embodiment of 'good mental health' is a disingenuous expectation. Critical scholarship has further demonstrated the entanglement of psychiatric practices in colonial projects, for example psychiatry's role in justifying the oppression of First Nations people, concealing First Nations knowledges, and perpetuating harm through coercive mental health practices (Dudgeon et al., 2017; Joseph, 2015). Nevertheless, a linear perspective on mental health progress (Bentall, 2011), which offers a comforting narrative of positive reform over time, means that it can be difficult to step back from the current moment in order to explore the limitations of contemporary psychiatric discourses and 'mental health' practices more generally. This chapter therefore seeks to explore how we might understand, research, and teach 'mental health' differently, by asking critical questions, holding the notion of 'mental health' more lightly, considering the layers of power involved in mental health service provision, and challenging the ways in which narrow forms of expertise marginalise experiential knowledges and obscure the politics of mental health.

Biomedical and Therapeutic Dominance

It is well-documented that medicalised perspectives on distress, predicated on interventions aimed at 'fixing' so-called dysfunctions within individuals, often do not meet people's needs during times of crisis and distress (Cosgrove & Shaughnessy, 2020). A biomedical paradigm can lead, for example, to practices that include diagnosing and medicating distress arising in the context of gendered violence, while failing to address material needs and safety (Humphreys & Thiara, 2003; Sweeney et al., 2019; Tseris et al., 2022). Rather than humility about the limitations of medical expertise in responding to complex human emotions, violence, and oppression, mental health professionals may express derogatory attitudes in response to distress that is not readily addressed via medication or that does not neatly fit into a diagnostic category (Lamont & Dickens, 2019). In addition, biomedical interventions are embedded in coercion and paternalism—in both overt and more subtle forms

(Campbell & Davidson, 2009); such practices can lead to further disempowerment and despair, rather than relief and validation (Berring & Georgaca, 2023).

There are vested interests that shape knowledge-power hierarchies (Williams & Mumtaz, 2008) in Western mental health systems. The commercial interests underpinning psychiatric research and pharmacological treatment guidelines have been rightly critiqued for distorting the evidence-base, despite ongoing claims of neutrality (Cosgrove & Vaswani, 2019). At the same time, it is striking how little attention has been given to the professional interests of human service professionals more broadly, in driving an expansion in *therapeutic* responses to distress (Moloney, 2013). Perhaps this is due to a misplaced assumption that therapy offers a much-needed, benign alternative to medication. An assumption about therapeutic benevolence, however, fails to engage with the neoliberal, patriarchal, and colonial underpinnings of mainstream therapy (Brown, 2021). It is, indeed, in the interests of human service professionals for therapy to thrive and be viewed as an inherently benevolent and effective response to distress. As a result of the substantial professional interests that shape dominant understandings of what is needed during times of crisis and suffering, user-led and survivor-led approaches to understanding human experience that critique and resist the centring of professionalised and medicalised assumptions (Beresford, 2021) may be sidelined and viewed sceptically by mental health professionals and, indeed, the general public. The idea that psychiatrised people might not 'need' professionals is a hazard to the continued growth of professional knowledges.

We are repeatedly told that we need to 'talk more' about mental health (Health Direct, 2024), but this assumes the relevance of medicalised and therapeutic mental health discourses. As a result, expertise on complex human experiences is placed into the hands of people trained in a few narrow fields (Faulkner, 2017). My hope as a lecturer and researcher is to invite some circumspection about what it is we might think that mental health professionals can offer, and to consider the value of engaging with more expansive knowledges for meaning-making during times of distress, crisis, or madness. Initially, this can be experienced as frustrating by some students, who are expecting to receive a particular form of knowledge that can inform 'professional' practice. A consideration of alternative paradigms can feel unusual, misguided, or even frightening. Yet, turning away from the limitations of psy-knowledges requires us to ignore the impacts of the biomedical, Eurocentric, and patriarchal logics underpinning professional expertise, which seek to define 'the problem' and then rush in paternalistically, unreflexively, and sometimes violently, to 'fix it', without consideration of the socio-political and cultural contexts, ethical frameworks, or limited worldviews that might be hindering our imaginations and actions.

Substantial work is therefore needed to disrupt the comfort, certainty, and perceived relevance of professional knowledges and paradigms that have systematically excluded and de-valued the knowledges of people with direct experiences of the mental health system (Beresford, 2021). This includes an over-reliance on rigid practice guidelines, checklists, and screening tools as a basis for understanding and responding to distress and crisis. It is heartening when students are able to challenge the entrenched idea of their future identity as a 'benevolent' professional who will help 'unwell others'. This is not easy learning to engage with, as questioning the mental health 'crisis' can involve a complex process of unlearning (Tseris & Jamieson, 2024). It is made even more challenging due to an entrenched perspective that it is an inherently benevolent act to intervene and 'correct' any thoughts, emotions, and behaviours that deviate from 'normal' and 'productive' citizenship (Esposito & Perez, 2014). Professional intervention is most often viewed as both a necessary and compassionate response to distress, as seen in mainstream mental health 'awareness-raising' campaigns and in mental health journalism (Holland, 2018). Consequently, alternatives to the current system are frequently obscured (Gee et al., 2016), including those led by psychiatric survivors. As this chapter discusses below, such power relations have led to substantial limitations on the scope and potential of mental health critique and 'reform' efforts as well.

Mental Health: Becoming Social?

Having outlined the persistence of biomedical and therapeutic knowledges as a way of making sense of human experience within Western contexts, it would be remiss to ignore the significant emergence of 'social' perspectives, leading to different possibilities for talking about 'mental health'. For example, although the biopsychosocial model has been rightly critiqued for acting more like the bio-bio-bio model (Whitley et al., 2015), it nonetheless has become an incredibly widespread paradigm in mental health services. Thus, notwithstanding its major limitations, 'the social' is a concept that has well and truly entered mental health contexts. Indeed, it is joined by a range of additional, newer frameworks—person-centred, strengths-based, recovery-oriented, trauma-informed care, among others! Such rhetoric has become commonplace within mental health services and is often held up as evidence that the mental health system is engaging in substantial and meaningful transformation (Becker-Blease, 2017)—pressing on forwards beyond the limitations of an individualistic lens, moving past pathologising and disempowering perspectives, towards a holistic and more 'progressive' approach.

Unfortunately, there is mounting evidence that the lived experiences of psychiatric survivors interfacing with mental health services often do not match the shiny rhetoric (Sweeney & Taggart, 2018; Tseris, 2024). It has been noted that mental health professionals can use the veneer of progress to conceal their ongoing commitments to coercive practices (Bentall, 2011). Indeed, while the promise of emancipatory approaches may offer promise of change and reform in the mental health system, sometimes 'new' ways of doing things simply reproduce existing power relations, co-opting the knowledge of psychiatric survivors into the existing system, as "activist concepts such as recovery, inclusion, access and hope have been co-opted, appropriated and politically neutralised by policy-makers, service-providers and government" (McWade et al., 2015, p. 307).

Consequently, any 'new' commitment made by mental health services to adopting 'progressive', 'socially-oriented', and 'lived experience-informed' perspectives must be viewed critically. An in-depth exploration of the professional interests shaping mental health reform efforts, potentially leading to a curtailing of the disruptive and transformative knowledges of mental health service users and psychiatric survivors, is therefore relevant.

'Lived experience' and the Hazards of Tokenism

It is now quite widely accepted that qualitative information is required to explore the complexities of mental distress and to explore first-person narratives of mental health system experiences (Cameron & McGowan, 2013). This is a welcome paradigm shift away from the systematic marginalisation and silencing of psychiatric survivors within the academy. As a result of the turn towards valuing first-person accounts of mental distress, contemporary approaches to mental health research and education have moved towards an increased emphasis on the importance of experiential knowledge (Faulkner, 2017). This shift holds promise in addressing the long-standing oppression of psychiatrised people, through situating people who have been diagnosed with mental illnesses as 'valid knowers' (Faulkner, 2021) rather than as 'passive recipients' of services.

However, it is not enough for the mental health system to try to 'listen to people with lived experience', nor to merely incorporate experiential expertise into mainstream ways of understanding and responding to distress. Critical scholarship has highlighted tokenism and micro-aggressions that are experienced by people contributing to mental health system reforms in 'peer' roles (Kilpatrick et al., 2017). Despite investment in a 'lived experience' workforce, transformative paradigm shifts in mental health systems remain rather elusive, as biomedical and therapeutic approaches retain their dominance and power (Gee et al., 2016). The obstacles to meaningful mental health reform are often discussed in neutral terms: they are described in terms of barriers, inadequate resources, conflicts between different points of view, or misunderstandings. Such neutral terms conceal the professional vested interests that govern the kinds of reform agendas that are pursued in mental health systems, the kinds of reform that are dismissed or considered impossible, and whose voices are invited to the table and whose are excluded (Daya et al., 2020). Psychiatric survivors have repeatedly outlined the unacceptable power imbalances and epistemic injustices that they face in attempting to work with mainstream services towards justice and transformation (Okoroji et al., 2023). This elucidates the limits that are placed around 'acceptable' knowledges and the ongoing framing of issues according to professional templates, despite new commitments to 'lived experience'. Thus, while critical theorists have rightly pointed to the interconnectedness between language, knowledge, and power in mental health, it does not necessarily follow that changing the language will automatically transform power dynamics and knowledge production in mental health.

Consequently, there is perhaps a little too much optimism about the power of language in driving reform and transformation. For example, it is well-documented that the concept of 'recovery' has been appropriated by mental health services to such an extent that many people do not recall its original intentions relating to a rejection of mainstream service provision (Sinclair et al., 2023). Perversely, the 'recovery' concept is sometimes invoked by mental health services to obscure ongoing coercive practices, for example, by labelling the removal of rights as 'supporting recovery' (Recovery in the Bin, 2017). Here, we can see that a shift in language can simply be a way of softening or sanitising a status quo approach—a concession which requires no concrete actions and in fact can obscure ongoing oppression. Its usage does not guarantee a critical approach or transformation of power arrangements in mental health services.

The Co-option of 'Trauma-informed Care'

In a contemporary context, it is increasingly common for trauma-informed practice to be invoked as evidence of 'genuine' mental health reform. Trauma-informed care has been heralded as a core ingredient of mental health system change and as a way of reducing, or even eliminating, adverse experiences in mental health systems (Becker-Blease, 2017). Although trauma-informed practice promises substantial and meaningful transformation within mental health services, there are many reasons to be cautious, as noted by Thompson (2021, p. 100), as "mainstream psychological accounts of trauma continue to draw on predominantly individualized and medicalized frameworks, couched in the clinical diagnostic language of the psy disciplines".

It is therefore important to remember that although trauma theory is not entirely unhelpful or irrelevant to mental health reform efforts, it certainly does not guarantee a transformative paradigm. Through the mainstreaming of the concept of trauma, its politicising and feminist roots have been largely subsumed into a therapeutic concept and practice 'tool' (Tseris, 2019). This is comforting for people with professional identities who wish to invoke notions of social justice, without meaningfully de-centring their 'expertise'. In practice, this can mean that professional power and the role of mental health 'experts' is potentially reinforced, rather than subverted, through trauma theory. Thus, the hazard is that we are left with a reinforced belief in the primacy of a psychological assessment of 'traumatised' individuals, while contextual factors are sidelined. There are additional implications of the contemporary turn to 'trauma-informed practices':

- Linearity—the Eurocentric idea that all distress is able to be explained and reduced to a single or straightforward cause, ignoring complexity and multiplicity, and centring the role of 'experts' in the assessment of trauma and individual 'recovery' (Thompson, 2021);
- Pathology—a focus on deficits, 'symptoms', and major 'mental health difficulties' among survivors of violence and abuse (Streeck-Fischer & van der Kolk, 2000);
- De-politicisation—meaning that we become so preoccupied with the idea of 'fixing people' within a psychological framing of trauma, that we risk losing an ability to recognise and respond to the violence of poverty, colonialism, racism, ableism, heteronormativity, and rape culture (Reynolds, 2020).

Stigma or Sanism?

The above examples elucidate the hazards of an 'add-on' approach to mental health reform, in which there has been a *sprinkling* of social perspectives onto biomedical and therapeutic templates of mental health. Such practices can perhaps be seen as micro-concessions, in response to decades of powerful critique and organising by the psychiatric survivor movement. However, concessions like these have often resulted in only superficial changes that have not fully recognised, let alone addressed, the lived experience of psychiatric harm, thus supporting the continuation of the mainstream professional project. Consequently, it is essential to engage with the limitations of changing words or language, or introducing new concepts or practice guidelines, without adequately disrupting status quo arrangements of power. The concept of epistemic injustice (Fricker, 2007) goes much further, requiring not only a change in the language used by professionals but a recognition of the multiple ways in which psychiatrised people's knowledges are silenced and discredited, leading to violent practices (Daya, 2022). In contrast, overly simplistic 'practice guidelines' may remain underpinned by notions of intra-individual pathology and professional benevolence (Cosgrove & Shaughnessy, 2020). Such approaches may offer neat 'truths' about human experience, which can seem highly compelling in an age of neoliberal higher education (Morley et al., 2017), especially for people with professional identities seeking knowledge that can be readily applied—and indeed commodified—in practice. This comforting knowledge differs substantially from a structural analysis of power relations in the mental health system.

Perhaps an additional component that is stifling more imaginative and transformative reform is the socialisation of mental health workers into a view that violence in mental health treatment is a largely historical phenomenon. This perspective encourages the view that while there are remnants of 'stigma' in the present-day, there is little ongoing institutional violence. However, 'stigma' is a notion that does not go nearly far enough in exploring the oppression of people in the mental health system. In fact, the idea of 'stigma' keeps the medical model intact, through a focus on problematic community attitudes, which distracts us from larger questions about how professional understandings and responses to distress justify and allow devastating experiences of psychiatric oppression in the present-day, including seclusion, restraint, and the informal coercion that underpins nearly every mental health intervention (Campbell & Davidson, 2009). In this way, the significant energy spent on talking about 'stigma' diverts our focus away from the harms of contemporary biomedical and therapeutic power. Poole et al. (2012) discuss the reasons why *sanism* is a much more useful concept, in contrast to the amorphous and somewhat de-politicising concept of 'stigma'. Sanism refers to "the systematic subjugation of people who have received 'mental health' diagnoses or treatment" (Poole et al., 2012, p. 20); importantly, the authors document examples of sanism that are widely accepted, everyday 'professional' practices in the contemporary mental health field. In this way, sanism provides space for exploring the oppressions involved in psychiatric labelling practices, for example the social exclusions arising from diagnoses and mental health treatment, including exclusion from employment and education opportunities. Sanism is also a concept that interrogates and disrupts taken-for-granted medical paternalism (Blanchette, 2019), which justifies the denial of psychiatrised people's dignity, autonomy, and liberty.

It is perhaps unsurprising that the concept of sanism has not yet gained significant traction in university curricula. Its inclusion would require an analysis of violence in mental health settings and the involvement (and often complicity) of all mental health professionals in oppressive practices, regardless of the articulation of seemingly empowering practice frameworks, including 'trauma-informed care'. Interestingly, sociology—which is potentially less invested in the professional 'mental health' project—has also shown a reluctance to interrogate the limitations of diagnostic paradigms, the harms of biomedical and therapeutic interventions, and an analysis of the connections between dominant constructions of 'mental health' and neoliberal society (Cohen, 2018); perhaps such reluctance can be linked to the demands of funding bodies and the relentless pressure on academics to demonstrate the alignment of our research with 'industry' needs.

A refusal to recognise and address the violence enacted within mental health settings is evident in small-scale reform agendas in mental health, which go some way towards acknowledging that there are 'problems' or 'flaws' to be solved in the mental health system, but which continue to justify the use of coercion, and fail to meaningfully address the social drivers of emotional suffering (Daley et al., 2019). Turning away from psychiatric violence justifies a continued denial of resources for peer-led and social justice-oriented responses, alongside continued energy and resources poured into bolstering the current system and reproducing the status quo within paradigms that claim to be transformative.

Moving Beyond a 'Mental Health' Framework

The inequalities of neoliberalism are stark. It is in the interests of those in power to situate people experiencing violence, poverty, and marginalisation as 'maladapted', through the language of psychiatric diagnosis. It is convenient to encourage the majority of us to become very focused on attaining 'good mental health', in order to deflect from more political and disruptive conversations (Cohen, 2016). It also makes sense that 'resilience' has become a highly desirable trait in contemporary times—as a supposedly successful and productive way of being in the world—because this places the onus on those with less power to accommodate to injustices and violence (Esposito & Perez, 2014). Within the university sector, there are immense pressures to accept and reproduce biomedical and therapeutic 'mental health' discourses in both research and teaching work. Other core institutions in the Australian context, including workplaces and schools, disseminate awareness-raising campaigns that compel us to persistently think and talk about our 'mental health'—although only in highly prescribed ways that reinforce biomedical and therapeutic governance (Holland, 2018). As a result, we risk losing our ability to account for the implications of injustices and unfair social conditions on people and communities, instead setting our gaze upon individuals who are 'failing to cope'. The positive messaging that surrounds the Western mental health system renders sociological critique quite complicated, as the social justice issue that receives the most attention is about 'increasing access' and 'scaling up' (Mills & Davar, 2016) the mental health system, rather than a questioning of the system's assumptions and limitations.

Despite the expansion of psy-discourses, there has also been significant, long-standing resistance to their universalising claims (Burstow, 2014). People with direct experiences of mental health systems have mounted incisive critiques in response to the limitations of biological determinism and neoliberal assumptions about 'ideal' personhood (Beresford, 2021). These knowledges provide important challenges to the professional mental health project, and it is a strategic move for mental health systems to not remain static in response to such critique but to attempt to appropriate and subsume experiential knowledges, in ways that retain and reinforce professional power (Recovery in the Bin, 2017). This chapter has explored how mental health services have navigated the call for more 'socially-oriented' understandings and responses, for example, in the examples of 'recovery-oriented practice' and 'trauma-informed care'. It is evident in these examples that mental health services have often failed to respond wholeheartedly to serious concerns about abuses of power in mental health systems and the need for meaningful paradigm shifts. As documented by Rose and Rose (2023, p. 47), it is troubling that "new ideas that seemed to have potential for a radical transformation of psychiatry, that is to say a shift in the power relations between professionals and users … seem doomed to be recuperated back into individualised treatment".

In addition to concerted critiques of the vested corporate interests shaping the pharmaceutical industry's relationship with psychiatric knowledge production (Cosgrove & Wheeler, 2013), we need to account for the role of professional vested interests in continuing long-established patterns of pathologising and individualising distress and madness, even within therapeutic and seemingly 'progressive' or 'socially-informed' mental health practices. Therefore, it is necessary to re-consider 'common sense' understandings of emotional suffering in order to re-conceptualise what is needed during times of crisis. This can create opportunities to engage with knowledges and actions that are rendered invisible when we only view distress through a diagnostic and individualising lens—ignoring that we are living in times of ecological destruction, war, devastating injustices, white supremacy, and gendered violence. From here, we might be able to better contextualise experiences of despair and loss, and to consider the breath of perspectives and knowledges that might be needed. To provide just one example, communities around the world are increasingly concerned by and impacted upon by human-induced climate change (Teasdale & Panegyres, 2023), but is such despair best understood as a 'mental health issue'? Through opening up different perspectives outside the normative 'mental health' paradigm, possibilities other than treating 'distressed' individuals can become clearer, and this has the potential to lead to creative forms of activism and collective responses (Corbett, 2017).

The priorities of professional training have been strongly critiqued on the basis that they usually do not include any reference to psychiatric survivor liberation movements, Mad histories, and anti-sanist practices (Poole et al., 2012). More than an innocent oversight, the medical, therapeutic, neoliberal, and carceral frameworks that underpin professional training systematically marginalise experiential knowledge, obscuring institutional violence, and precluding a socio-political analysis of mental distress and the mental health system (Cosgrove & Shaughnessy, 2020). The vested professional interests that each mental health discipline has in increasing its power, and in maximising its perceived relevance in responding to mental distress, is a clear factor in understanding this silence. This creates powerful drivers for maintaining and, indeed, growing mainstream mental health services, to produce an ever-expanding professional project that is, at best, ill-suited to addressing contexts of injustice and systemic oppression, and that often exacerbates despair (Whitaker et al., 2021). The dominance of such perspectives denies people and communities the right to be respected for their own local knowledges, rather than being pathologised for refusing unwanted forms of 'help' (Fortier & Wong, 2019).

Nevertheless, everyday dissent and resistance against oppressive and neoliberal frameworks and narrow professional agendas demonstrates their fragility and allows alternative discourses to emerge, including the ethics of interdependency, mutual caring, and solidarity (Garrett, 2021). Herein lies the importance of moving beyond a 'social' perspective that acts as a mere 'add-on' to the status quo. Instead, we must make space for a critique of the colonial, neoliberal, and heteropatriarchal assumptions that underpin medical and therapeutic models of mental health. This is quite challenging, as it necessitates moving into a place of uncertainty, and branching out beyond established parameters, in order "not to fix a broken system, but to acknowledge our inherent power to summon other worlds" (Akomolafe & Benavides, 2022, n.p.). Indeed, research and teaching in the field of mental health ought to be engaging in a broad array of transformative and dissenting discourses. However, the neoliberalisation of universities is a huge barrier that needs to be destabilised, in order to make space for more expansive and liberatory research and teaching agendas (Morley et al., 2017).

Within university settings, a good start might be to re-consider how the resources of the academy ought to be used. When we consider the extent of structural and institutional violence in psychiatrised people's lives (Daley et al., 2019), it becomes increasingly difficult to justify the enormous resources that continue to be allocated to diagnostic and individual-focused approaches, especially when we acknowledge how such approaches frequently enact further oppression and silencing practices (Sweeney et al., 2019). Countering psychiatric harm requires a transformative paradigm shift, away from centring narrow forms of professional expertise, and this will not be achieved through integration of mildly 'social' perspectives into the mainstream mental health system, nor modest tweaks in the mental health lexicon.

Attention is needed on the inequalities that produce and exacerbate experiences of distress, on user/survivor activism focused on ending psychiatric harm and sanism, and on broader social movements aimed at ending a range of intersecting oppressions, including cisheteronormativity, poverty, settler colonialism, and white supremacy. There is already substantial work being done to challenge the silencing of psychiatric survivors within the academy. Mad Studies courses are creating exciting new parameters for tertiary education, away from training 'experts' and towards teaching "mad people's histories, cultures, politics, and communities … and use mad-centred knowledge to critique existing cultures and practices" (Spandler & Poursanidou, 2019, p. 9). People with professional identities and training can learn from, engage in, and contribute to survivor-led alternatives, if they are able to acknowledge that such actions and alliances require humility and a joining-with rather than power-over approach (Beresford, 2021). As noted by Indigo Daya (2022, p. 11), an Australian mad survivor activist, researcher, and artist, "genuine allies are welcomed and much needed. However, the risks of epistemic injustice will remain if we do not shift from the position that traditional academia is best placed to interpret lived experience".

The need to acknowledge and end the harms of psychiatric coercion should be core reading for all students engaging with the mental health field, regardless of their discipline. There are valuable resources to support this learning, both international scholarship (e.g. Indregard et al., 2024), and resources produced in Australia, including by people with lived experiences of involuntary mental health treatment (e.g. Gooding et al., 2018; Katterl et al., 2023). Through de-centring biomedical and therapeutic power, the narrow set of knowledges produced by psy-disciplines can be questioned, and the assertion that human service professionals own all the relevant expertise to respond to despair, crisis, and suffering, can be disrupted.

References

Akomolafe, B., & Benavides, M. (2022). The times are urgent: Let's slow down. https://www.bayoakomolafe.net/post/the-times-are-urgent-lets-slow-down

Becker-Blease, K. A. (2017). As the world becomes trauma–informed, work to do. *Journal of Trauma & Dissociation, 18*(2), 131–138. https://doi.org/10.1080/15299732.2017.1253401

Bentall, R. P. (2011). *Doctoring the mind: Why psychiatric treatments fail.* Penguin Books.

Beresford, P. (2021). Introduction. In P. Beresford & J. Russo (Eds.), *The Routledge international handbook of mad studies* (pp. 1–16). Routledge.

Berring, L. L., & Georgaca, E. (2023). A call for transformation: Moving away from coercive measures in mental health care. *Healthcare, 11*(16). https://doi.org/10.3390/healthcare11162315

Blanchette, S. (2019). A feminist bioethical and mad studies approach to resisting an increase in psychiatric paternalism to competent mental health users/refusers. *Journal of Ethics in Mental Health, 6*(10), 1–18.

Breen, L., & Darlaston-Jones, D. (2008). Moving beyond the enduring dominance of positivism in psychological research: An Australian perspective.

43rd Australian Psychological Society Annual Conference. https://researchonline.nd.edu.au/cgi/viewcontent.cgi?article=1002&context=arts_conference

Brown, C. (2021). Critical clinical social work and the neoliberal constraints on social justice in mental health. *Research on Social Work Practice, 31*(6), 644–652. https://doi.org/10.1177/1049731520984531

Burstow, B. (2014). The withering away of psychiatry: An attrition model for antipsychiatry. In B. Burstow, B. LeFrancois, & S. Diamond (Eds.), *Psychiatry disrupted. Theorizing resistance and crafting the (r)evolution* (pp. 34–51). McGill-Queen's University Press.

Cameron, D., & McGowan, P. (2013). The mental health social worker as a transitional participant: Actively listening to 'voices' and getting into the recovery position. *Journal of Social Work Practice, 27*(1), 21–32. https://doi.org/10.1080/02650533.2012.732047

Campbell, J., & Davidson, G. (2009). Coercion in the community: A situated approach to the examination of ethical challenges for mental health social workers. *Ethics and Social Welfare, 3*(3), 249–263. https://doi.org/10.1080/17496530903209469

Cohen, B. M. (2016). *Psychiatric hegemony: A Marxist theory of mental illness*. Springer.

Cohen, B. M. (2018). Introduction. In B. M. Cohen (Ed.), *Routledge international handbook of critical mental health* (pp. 1–12). Routledge.

Corbett, S. P. (2017). *How to be a craftivist*. Unbound Publishing.

Cosgrove, L., & Shaughnessy, A. F. (2020). Mental health as a basic human right and the interference of commercialized science. *Health and Human Rights, 22*(1), 61–68.

Cosgrove, L., & Vaswani, A. (2019). The influence of pharmaceutical companies and restoring integrity to psychiatric research and practice. In S. Steingard (Ed.), *Critical psychiatry: Controversies and clinical implications* (pp. 71–96). Springer.

Cosgrove, L., & Wheeler, E. E. (2013). Industry's colonization of psychiatry: Ethical and practical implications of financial conflicts of interest in the DSM-5. *Feminism & Psychology, 23*(1), 93–106. https://doi.org/10.1177/0959353512467972

Daley, A., Costa, L., & Beresford, P. (2019). Introduction. In A. Daley, L. Costa, & P. Beresford (Eds.), *Madness, violence, and power: A critical collection* (pp. 3–20). University of Toronto Press.

Davies, J. (2021). *Sedated: How modern capitalism created our mental health crisis*. Atlantic Books.

Daya, I. (2022). Russian dolls and epistemic crypts: A lived experience reflection on epistemic injustice and psychiatric confinement. *Incarceration, 3*(2). https://doi.org/10.1177/26326663221103445

Daya, I., Hamilton, B., & Roper, C. (2020). Authentic engagement: A conceptual model for welcoming diverse and challenging consumer and survivor views in mental health research, policy, and practice. *International Journal of Mental Health Nursing, 29*(2), 299–311. https://doi.org/10.1111/inm.12653

Department of Health and Aged Care. (2022). *About mental health*. https://www.health.gov.au/topics/mental-health-and-suicide-prevention/about-mental-health

Dudgeon, P., Calma, T., & Holland, C. (2017). The context and causes of the suicide of Indigenous people in Australia. *Journal of Indigenous Wellbeing, 2*(2), 5–15.

Enticott, J., Dawadi, S., Shawyer, F., Inder, B., Fossey, E., Teede, H., et al. (2022). Mental health in Australia: Psychological distress reported in six consecutive cross-sectional national surveys from 2001 to 2018. *Frontiers in Psychiatry, 13*, 815904. https://doi.org/10.3389/fpsyt.2022.815904

Esposito, L., & Perez, F. M. (2014). Neoliberalism and the commodification of mental health. *Humanity & Society, 38*(4), 414–442. https://doi.org/10.1177/0160597614544958

Faulkner, A. (2017). Survivor research and Mad Studies: The role and value of experiential knowledge in mental health research. *Disability & Society, 32*(4), 500–520.

Faulkner, A. (2021). Knowing our own minds: Transforming the knowledge base of madness and distress. In R. Ellis, S. Kendal, & S. Taylor (Eds.), *Voices in the history of Madness: Personal and professional perspectives on mental health and illness* (pp. 333–357). Springer International Publishing.

Fortier, C., & Wong, E. H. S. (2019). The settler colonialism of social work and the social work of settler colonialism. *Settler Colonial Studies, 9*(4), 437–456. https://doi.org/10.1080/2201473x.2018.1519962

Fricker, M. (2007). *Epistemic injustice: Power and the ethics of knowing*. Oxford University Press.

Gao, C. X., McDonald, L. P., Hamilton, M. P., Simons, K., Menssink, J. M., Filia, K., et al. (2023). Inequalities in access to mental health treatment by Australian youths during the COVID-19 pandemic. *Psychiatric Services, 74*(6), 581–588. https://doi.org/10.1176/appi.ps.20220345

Garrett, M. P. (2021). 'A world to win': In defence of (dissenting) social work – A response to Chris Maylea. *The British Journal of Social Work, 51*(4), 1131–1149. https://doi.org/10.1093/bjsw/bcab009

Gee, A., McGarty, C., & Banfield, M. (2016). Barriers to genuine consumer and carer participation from the perspectives of Australian systemic mental health advocates. *Journal of Mental Health, 25*(3), 231–237. https://doi.org/10.3109/09638237.2015.1124383

Gooding, P., McSherry, B., Roper, C., & Grey, F. (2018). *Alternatives to coercion in mental health settings: A literature review*. Melbourne Social Equity Institute, University of Melbourne.

Health Direct. (2024). *How to talk about your mental health concerns*. https://healthdirect.gov.au/how-to-talk-about-your-mental-health-concerns#:~:text=Don't/20disclose/20everything/20at,it/20out/20loud/20to/20yourself

Holland, K. (2018). Lay theories and criticisms of mental health news: Elaborating the concept of biocommunicability. *Disability & Society, 33*(8), 1327–1348. https://doi.org/10.1080/09687599.2018.1487831

Humphreys, C., & Thiara, R. (2003). Mental health and domestic violence: 'I call it symptoms of abuse'. *British Journal of Social Work, 33*(2), 209–226. https://doi.org/10.1093/bjsw/33.2.209

Indregard, A. M. R., Nussle, H. M., Hagen, M., Vandvik, P. O., Tesli, M., Gather, J., & Kunøe, N. (2024). Open-door policy versus treatment-as-usual in urban psychiatric inpatient wards: A pragmatic, randomised controlled, non-inferiority trial in Norway. *The Lancet Psychiatry*. https://doi.org/10.1016/s2215-0366(24)00039-7

Joseph, A. J. (2015). The necessity of an attention to Eurocentrism and colonial technologies: An addition to critical mental health literature. *Disability & Society, 30*(7), 1021–1041. https://doi.org/10.1080/09687599.2015.1067187

Katterl, S., Lambert, C., MacBean, C., Grey, F., Downes, L., Cataldo, M., Clarke, K. & Williams, S. (2023). *Not before time: Lived experience justice and repair*. https://www.livedexperiencejustice.au/

Kilpatrick, E., Keeney, S., & McCauley, C. O. (2017). Tokenistic or genuinely effective? Exploring the views of voluntary sector staff regarding the emerging peer support worker role in mental health. *Journal of Psychiatric and Mental Health Nursing, 24*(7), 503–512. https://doi.org/10.1111/jpm.12391

Lamont, E., & Dickens, G. L. (2019). Mental health services, care provision, and professional support for people diagnosed with borderline personality disorder: Systematic review of service-user, family, and carer perspectives. *Journal of Mental Health, 5*, 619–633. https://doi.org/10.1080/09638237.2019.1608923

McWade, B., Milton, D., & Beresford, P. (2015). Mad studies and neurodiversity: A dialogue. *Disability & Society, 30*(2), 305–309. https://doi.org/10.1080/09687599.2014.1000512

Mills, C., & Davar, B. (2016). A local critique of global mental health. In S. Grech & K. Soldatic (Eds.), *Disability in the Global South: The critical handbook* (pp. 437–451). Springer.

Moloney, P. (2013). *The therapy industry: The irresistible rise of the talking cure, and why it doesn't work.* Pluto Press.

Morley, C., Macfarlane, S., & Ablett, P. (2017). The neoliberal colonisation of social work education: A critical analysis and practices for resistance. *Advances in Social Work and Welfare Education, 19*(2), 25–40.

Okoroji, C., Mackay, T., Robotham, D., Beckford, D., & Pinfold, V. (2023). Epistemic injustice and mental health research: A pragmatic approach to working with lived experience expertise. *Frontiers in Psychiatry, 14*, 1114725. https://doi.org/10.3389/fpsyt.2023.1114725

Poole, J. M., Jivraj, T., Arslanian, A., Bellows, K., Chiasson, S., Hakimy, H., et al. (2012). Sanism, 'mental health', and social work/education: A review and call to action. *Intersectionalities: A Global Journal of Social Work Analysis. Research, Polity, and Practice, 1*, 20–36. https://doi.org/10.32920/21751490.v1

Recovery in the Bin. (2017). *RITB – Key principles.* https://recoveryinthebin.org/ritbkeyprinciples/

Reynolds, V. (2020). Trauma and resistance: 'Hang time' and other innovative responses to oppression, violence and suffering. *Journal of Family Therapy, 42*(3), 347–364. https://doi.org/10.1111/1467-6427.12293

Rose, D., & Rose, N. (2023). Is 'another' psychiatry possible? *Psychological Medicine, 53*(1), 46–54. https://doi.org/10.1017/s003329172200383x

Rose, N. (2008). Psychology as a social science. *Subjectivity, 25*, 446–462. https://doi.org/10.1057/sub.2008.30

Sinclair, A., Gillieatt, S., Fernandes, C., & Mahboub, L. (2023). Inclusion as assimilation, integration, or co-optation? A post-structural analysis of inclusion as produced through mental health research on peer support. *Qualitative Health Research, 33*(6), 543–555. https://doi.org/10.1177/10497323231163735

Spandler, H., & Poursanidou, D. (2019). Who is included in the Mad Studies Project? *The Journal of Ethics in Mental Health, 10.*

Streeck-Fischer, A., & van der Kolk, B. A. (2000). Down will come baby, cradle and all: Diagnostic and therapeutic implications of chronic trauma on child development. *Australian & New Zealand Journal of Psychiatry, 34*(6), 903–918. https://doi.org/10.1080/000486700265

Sweeney, A., Perôt, C., Callard, F., Adenden, V., Mantovani, N., & Goldsmith, L. (2019). Out of the silence: Towards grassroots and trauma-informed support for people who have experienced sexual violence and abuse. *Epidemiology and Psychiatric Sciences, 28*(6), 598–602. https://doi.org/10.1080/09638237.2018.1520973

Sweeney, A., & Taggart, D. (2018). (Mis) understanding trauma-informed approaches in mental health. *Journal of Mental Health, 27*(5), 383–387.

Teasdale, N., & Panegyres, P. K. (2023). Climate change in Western Australia and its impact on human health. *The Journal of Climate Change and Health, 12*, 100243.

Thompson, L. (2021). Toward a feminist psychological theory of "institutional trauma". *Feminism & Psychology, 31*(1), 99–118. https://doi.org/10.1177/0959353520968374

Tseris, E. (2019). *Trauma, women's mental health, and social justice: Pitfalls and possibilities*. Routledge.

Tseris, E. (2024). The myth of trauma-informed care. In C. Brown (Ed.), *Reframing trauma through social justice: Resisting the politics of mainstream trauma discourse* (pp. 57–70). Routledge.

Tseris, E., & Jamieson, S. (2024). Mental health and biomedical neoliberalism: Can creativity make space for social justice? *Social Work Education*. https://doi.org/10.1080/02615479.2024.2328790

Tseris, E. J., Bright Hart, E., & Franks, S. (2022). "My voice was discounted the whole way through": A gendered analysis of women's experiences of involuntary mental health treatment. *Affilia, 37*(4), 645–663. https://doi.org/10.1177/08861099221108714

Whitaker, L., Smith, F. L., Brasier, C., Petrakis, M., & Brophy, L. (2021). Engaging with transformative paradigms in mental health. *International Journal of Environmental Research and Public Health, 18*(18), 9504. https://doi.org/10.1503/cmaj.141558

Whitley, R., Palmer, V., & Gunn, J. (2015). Recovery from severe mental illness. *Canadian Medical Association Journal, 187*(13), 951–952. https://doi.org/10.1503/cmaj.141558

Williams, L., & Mumtaz, Z. (2008). Being alive well? Power-knowledge as a countervailing force to the realization of mental well-being for Canada's aboriginal young people. *International Journal of Mental Health Promotion, 10*(4), 21–31. https://doi.org/10.1080/14623730.2008.9721773

Wykes, T., Bell, A., Carr, S., Coldham, T., Gilbody, S., Hotopf, M., et al. (2023). Shared goals for mental health research: What, why and when for the 2020s. *Journal of Mental Health, 32*(6), 997–1005. https://doi.org/10.1080/09638237.2021.1898552

4

Educating Emotions for Well-Being in Aotearoa New Zealand

Roberto McLeay

Introduction

The mental health and well-being of young people continue to be an ongoing concern in Aotearoa New Zealand (Mental Health Foundation, 2023; Ryan, 2024). Recent studies have reported that children and youth are struggling with issues of depression, anxiety, suicide, and self-harm as the result of cyber abuse, bullying, climate change, and other post-pandemic living challenges (see McLeay et al., 2023 for further examples). Consequently, a $1.9 billion endeavour to address these severe consequences has led to various strategies being implemented in Aotearoa New Zealand, such as the enhancement of youth mental health services, provision of early interventions, and creation of a 'newer' mental health workforce (Mental Health & Wellbeing Commission, 2023; Ryan, 2024). Among these approaches, teaching individuals how to observe,

R. McLeay (✉)
The University of Auckland, Auckland, New Zealand
e-mail: roberto.mcleay@auckland.ac.nz

N. A. Hendry, E. Karageorgos (eds.), *Critical Mental Health in Australia and Aotearoa New Zealand*, The Politics of Mental Health and Illness,
https://doi.org/10.1007/978-3-031-82861-4_4

manage, and regulate their emotions has gained recognition as an effective strategy for helping young people with their well-being (Ministry of Education, 2023; The Education Hub, 2024b). Although teaching children about their feelings and how to control them appears to be 'progressive' and 'positive', this chapter raises concern for how uncritically educating emotions in schools may inspire young people to govern and relate themselves (as well as others) to problematic discourses promoted by psy-professionals.

The Discourse of Happiness and Well-Being: Driving the Need for a Caring Education

Today, well-being has come to be recognised as a universal and incontestable goal that individuals, families, communities, and nations should direct themselves towards (Binkley, 2014). Although well-being is now promoted by a plethora of individuals, professions, and industries (also see Davies, 2015), one particular field may be recognised as having a significantly influential role in its conceptualisation and rise to popularity: positive psychology.

Disputing the long-held view of emotions as pathologies that disrupt bio-organic functioning (American Psychiatric Association, 2013; Gorton, 1873), positive psychologists, such as Martin Seligman (2002), posited that feelings, especially *happiness*, served to support individuals with 'thriving' (Binkley, 2014) and living a 'good life' (Davies, 2015). While this repositioned emotions as desirable motivators (see Salovey & Mayer, 1990), the evolving discourse on well-being transformed the perception of happiness. No longer seen through its social-cultural origins—as a feeling of luck, chance, or success arising from largely unpredictable and uncontrollable circumstances (Watt Smith, 2015)—happiness began to be regarded in the twentieth century as a state that could be attained and maintained through deliberate effort and control (Cabanas, 2016). Consequently, well-being (and happiness) has come to be viewed as an imperative that is 'rightfully' pursued by both old and young people alike. The promotion of 'happiness' as well-being (and vice-versa) has also subtly and successfully authorised emotions to be discriminated as either 'positive/happy' or 'negative/unhappy' (Hyman, 2014). By establishing these binary categories, positive psychologists proposed that people could foster a 'strength of character' (Seligman, 2002), whereby finding ways to attain and maintain 'positive' emotions, while learning how to prevent or reduce 'negative' ones, could help individuals endure the challenges of life.[1] Here, happiness may be seen to shift from a collective responsibility to a project of individualisation and personal responsibility (Cabanas, 2016), as each and every individual becomes accountable for cultivating their own well-being and character (Binkley, 2014). Yet, the proliferation of studies, measures, services, and interventions focused on fostering happiness and related virtues (see Carr, 2011; Park et al., 2004) also illustrates how the goal to develop and maintain a character of well-being has become a significant collective concern for societies and communities.

While the drive to develop a 'positive character' appears beneficial, there are concerns regarding the discourse of well-being. Scholars argue that the process of shaping emotions, attitudes, and the temperament to be happy is a subtle political tactic designed to meet the demands of an increasingly capitalist, neoliberal workforce (Binkley, 2014; Cabanas & Illouz, 2019; Cohen, 2016), as the 'measurable' outcomes of children and young people's well-being often focus on enhancing school attendance, behaviour, productivity, and overall academic achievement (Child & Youth Wellbeing, 2019; McLeod & Wright, 2016; Ministry of Education, 2017, 2023). Although positive psychology heavily shaped the modern discourse of well-being (Hyman, 2014), the specific drive to care for children and young people's well-being draws attention to another psychological field that has concerned itself with the developing character of young people: developmental psychology.

Although developmental psychology is largely regarded as a scientific discipline (particularly given its longer history compared to positive psychology), its conception and continuation have relied on the philosophical assertion that physical matter is the primary, if not only, cause of

[1] While scholars have noted that negative emotions have also contributed to the development of positive psychology (see, e.g., Pauwels, 2015), happiness, positive character strengths, and positive emotions are central to positive psychology.

human existence (Rose, 1999). Hence, child development serves to transform the intangible social-cultural-political aspects of children and youth—their emotions, morality, thoughts, and conduct—into neutral, observable, measurable, and comparable features (Binkley, 2014). By privileging the well-being discourse and its specific belief in positive emotions for optimal human existence (Seligman, 2002), developmental psychology joined to support the naturalisation and normalisation movement of happiness (Cabanas, 2016) as the state of 'healthy' character for young people (Rose, 1999). Subsequently, children have come to be saddled with the responsibility and expectation to be 'normal', healthy, and happy (Binkley, 2014) or learn how to do so (see, e.g., McLeay et al., 2023). Against this frame of normalcy, young people were also able to be judged as 'unhealthy' and deemed as troubling, deviant, morally concerning individuals (Cohen, 2016; McLeay & Powell, 2022) and in need of various strategies that focus on re-forming, re-producing, and re-shaping their emotions to guide their personality's back to the 'norm' (Rose, 1999). However, the concern for developing children's identity is not new and reflects earlier ideas, such as mental hygiene that focused on helping individuals develop a 'positive' or 'wholesome' personality and soul (for more details see McLeay, 2025). The universal, moral, claim of the naturalness and healthiness of well-being is also contestable, as happiness, for example, was once considered a feeling of the foolish, while sadness was valued for fostering a steadfast, sound character in European societies during the Renaissance (Watt Smith, 2015). The positive idea of attaining health identity epitomises bio-power (Foucault, 1975), where individuals—who are shaped by the discourse of well-being—exercise themselves freely to strive for maximum happiness and, in doing so, become an increasingly manageable and governable population (Binkley, 2014; Duncan, 2019).

Although scholars have continued to highlight political, cultural, and social issues within the discourses of well-being and emotions—ranging from a lack of definitional clarity (Burkitt, 2018) and unreliable measures (Bond & Lang, 2019) to its colonial influence (Holmes, 2021)—the push from various psy-disciplines (Binkley, 2014) has led to a 'therapeutic turn' in society and education (see Wright, 2014). Consequently, educational legislation and guidelines have intensely focused on young people's emotions (also see McLeay & Powell, 2022) and directed schools, with their long-standing focus on educating public health and well-being imperatives (Gard & Pluim, 2014; McLeod & Wright, 2016), to turn to various psychological concepts, pedagogies, strategies, and learning activities in an attempt to effectively support and care for young people. The following section details one particular example that exemplifies the integration of well-being and this therapeutic shift in education: social emotional learning (SEL).

A Strategy for Educating Young People's Feelings: Social Emotional Learning

Social emotional learning (SEL) has become a popular global strategy for supporting young people, which involves whole school systems, programmes, classroom strategies, lessons, and resources (Collaborative for Academic, Social, and Emotional Learning, 2024b). According to United States-based leading international organisation Collaborative for Academic, Social, and Emotional Learning (CASEL) (2024a, para. 1), SEL is a complex process whereby individuals "[1] acquire and apply the knowledge, skills, and attitudes to [2] develop healthy identities, [3] manage emotions and [4] achieve personal and collective goals". Although this process appears 'positive' and 'progressive', the following sub-sections examine how SEL may serve as a subtle strategy of social control; *disciplining*—training and exercising (Foucault, 1995)—young people, through the dominant discourses of well-being, to privilege psy-professionals voices, views, and beliefs on their feelings. Although the aims of SEL from developmental psychology have been embraced and amplified by positive psychology initiatives, distinguishing the specific practices of each and their implementation in educational settings has become increasingly challenging to discern. Moreover, while the histories and endeavours of positive psychology, developmental psychology, and SEL are not entirely commensurate, each has contributed over time to legitimising and strengthening the others, with positive psychology focusing on SEL's engagement with emotional learning (as this chapter will later show).

Applying Knowledge, Skills, and Attitudes

SEL relies on a wide range of engaging activities and tasks to help young people develop knowledge, skills, and attitudes with their own and others' emotions. For example, encouraging children to play with "visuals and picture cards that show emotions" (The Education Hub, 2024c, para. 8) illustrates how SEL helps young people develop skills of *observing* human feelings. By focusing intentionally on visual images, this game *disciplines* young people to exercise the dominant idea that emotions are *known* as physical, biological, bodily expressions. However, scholars have criticised this understanding, arguing instead that emotions are socially and culturally shaped (Burkitt, 2018; McLeay & Powell, 2022; Watt Smith, 2015). Providing time in class for young people "to read the emotions of their peers and think about what they could do to support their peer when they are sad or lonely" (The Education Hub, 2024c, para. 9) illustrates the practical application of observation *skills* and demonstrates how students may strategically exercise the dominant discourse of well-being to direct themselves (as well as others) towards valuing certain feelings (like happiness or calmness) while showing concern for other feelings (such as anger or sadness). Consequently, children develop a fundamental SEL *attitude* that certain emotions are desirable and should be pursued, while others should be 'helped'.

Develop Healthy Identities

Through developing their *knowledge*, *skills*, and *attitudes* of SEL across all levels of education from early childhood to higher education, children and young people are encouraged to develop the understanding that there is a 'desired' healthy identity to be achieved. Here, the categories 'positive' and 'negative' may support young people to *discipline* themselves towards happiness, positivity, calmness, and other positive emotions that signify the identity of 'strength' and 'well-being' (also see McLeay et al., 2023). The concept of 'self' plays an important role in this responsibilising and individualising process (Hyman, 2014; Rose, 1999), as children and young people learn to take responsibility for the formation of their own personality development through learning how to manage their emotions.

Manage Emotions

The developing *knowledge*, *skills*, and *attitudes* of SEL also support young people to develop and understand that emotions can and must be managed. Although emotional management is a central pillar of the contemporary SEL process, CASEL co-founder and SEL expert Goleman (1995) attributed it to the early work of Emotional Intelligence (EI) psychologists Peter Salovey and John Mayer (1990), who described emotional management as part of a complex process that involved monitoring and discriminating feelings to "guide one's thinking and actions" (p. 190). Like the skill of observation, the focus on 'monitoring' and discrimination (Salovey & Mayer, 1990) illuminates the importance of *surveillance* (Foucault, 1991)—where individuals engage in observing feelings against a dominant discourse—for managing emotions. Managing emotions for both EI and SEL therefore involves helping individuals learn how to examine their emotions against the discourse of emotional well-being and act accordingly.

Achieving Individual and Collective Goals

Through the process of SEL children position the 'norms' of well-being as important goals to navigate themselves towards. Subsequently, young people develop personal goals of cultivating positive emotions (such as happiness and calmness) while mitigating negative ones (like sadness or fear) (see McLeay et al., 2023). This goal results in young people seeking professional counselling services and self-help tools, like digital apps (Walsh, 2024), to enhance their emotional well-being. While this illustrates the formation of individual goals, the personal aim of obtaining 'positive feelings' also supports the collective societal goal to produce young entrepreneurial citizens (Ministry of Education, 2007) who display attitudes and behaviours that support their attendance, work, academic achievements, and 'success' in school (Corcoran et al., 2018; Organisation for Economic Co-operation and Development, 2023). However, this goal to attain certain emotions has been argued as an ideal strategy of social control that produces individuals who will guide and govern themselves towards the desired character needed for a productive neoliberal society (Burkitt, 2018; Cabanas, 2016; Davies, 2015).

Despite the concerns and issues with well-being (McLeod & Wright, 2016), happiness (Duncan, 2019), and emotions (McLeay et al., 2023), the uncritical welcoming of positive psychology initiatives and SEL into schools across the globe (OECD, 2023) has led to this chapter carefully investigating how it is deployed and promoted within local education systems. The following section outlines Foucault's conceptualisation of governmentality, which is used as the theoretical lens for critically analysing and discussing the subsequent case study on educating emotions in Aotearoa New Zealand early childhood.

Foucault's Governmentality

The concept of governmentality emerged in the 1970s through Foucault's scholarship on political power, territory, security, government, and the development of 'the population'. Although Foucault had previously spoken about government (see Foucault, 1995, for example), his 1978 lecture at the Collège de France, titled 'Governmentality' (Foucault, 1991), served as a significant moment where he introduced the distinctive term. Here, Foucault invited scholars to develop an understanding of governmentality by presenting them with an 'inventory of questions' that the government had been concerned with since the sixteenth century: "[h]ow to govern oneself, how to be governed, how to govern others, by whom the people will accept being governed, how to become the best possible governor" (Foucault, 1991, p. 87). These questions not only illustrated the early 'problems of government' and its population but also highlighted how the development of government rationality to address these issues focused on finding ways to utilise the networks of relations between and within individuals (Foucault, 1996). Thus, rather than analysing 'government' as a specific individual or group who imposes their will onto others, Foucault (1991) promoted recognising government as an "ensemble" of authorities, institutions, and agents who employ a variety of technologies, tactics, and bodies of knowledge to guide the conduct of individuals towards certain, albeit unpredictable, ends (p. 102). When it comes to educating emotions, there is a vast array of actors—such as health and education experts (Collaborative for Academic, Social, and Emotional Learning, 2024a), government organisations (Ministry of Education, 2023), philanthropists (Bill & Melinda Gates Foundation, 2023), and psychology influencers (Shadé, 2022)—who take part in this 'ensemble' of government. By promoting certain concepts, programmes, and services that are designed to help young people learn how to observe, manage, and regulate their emotions, this 'ensemble' works to direct how individuals and populations know, see, and, therefore, enact their feelings.

For Foucault (1996, p. 410) conceptualising power as a "relationship between individuals" rather than "a thing [an object to be held, possessed, or removed]" is fundamental for understanding both government and governmentality. By taking this stance, he challenged the view that 'government' is only a restrictive or oppressive relation, where power is exercised from the top-down (Foucault, 1995), and opened up the idea that power could be dispersed in numerous directions—working itself within, from, and between every person in an intricate network of relationships (Foucault, 1980, 1991). Power, therefore, may be seen to be everywhere, every-present, and always at work (Foucault, 1996), as people are constantly in relations with each other as well as themselves. Individuals (such as young people, parents, or teachers), who pursue and encourage learning about emotions in order to control them, may therefore be identified as a part of the intricate 'ensemble' of government. This mounting network of governing power-relations should not be viewed as 'negative' or 'restrictive', but as Foucault (1996) explains: "is positive in its effect. Power invents, power creates, [and] power produces" (p. 158). Although educating young people's emotions uses the positive, productive, capacities of disciplinary and bio-power (also see McLeay & Powell, 2022), the following case study draws particular attention towards a lingering power that supports the creation of governable individuals and populations: pastoral power.

According to Foucault (1997, 2009), pastoral power originated from early religious practices and can be best understood through the metaphor of a 'shepherd', who uses a range of techniques and strategies to gather, monitor, resource, and guide large populations of individuals (the flock). Within a learning community, teachers, parents, principals, counsellors, and other educational experts may be seen to reflect this role of a 'shepherd' (Foucault, 1997). While shepherds skilfully utilise a wide range of interventions and practices to direct a flock, the effectiveness of their role is linked to their capacity to foster unique relationships characterised by virtues of care and responsibility (Foucault, 2009). Thus, rather than being negative or oppressive, the devoted shepherd uses a unique caring relationship—described by Foucault (2009) as the shepherd/flock relation—to exercise a positive, productive, and creative power that guides and shapes the flock. By granting "as much value to a single one of the sheep as to the entire flock" (Foucault, 1997, p. 68), the shepherd/flock relationship encourages the shepherd to focus on each and every individual to better grasp how to guide their innermost feelings, thoughts, and conduct (Foucault, 2009). Strategies of knowledge and control, such as confessions and examination (Foucault, 1991, 2009), are important to achieve this, as they lead individuals to learn how to reveal their inner, private lives. Consequently, these strategies also authorise shepherds, who manage to gain insight into every individual sheep, to better understand how to manage and guide the entire flock. The shepherd/flock relation therefore exercises an individualising and totalising power which allows for greater social management and control over young individuals and populations in Aotearoa New Zealand. While pastoral power may be traced back to religious contexts (Foucault, 2009), its caring tactics and techniques have gradually transformed and become embedded within secular institutions like schools (Foucault, 1997), which aim to move young people towards their "salvation" (Foucault, 1997) or, in this case, emotional well-being.

Case Study: Emotional Learning in Aotearoa New Zealand Early Childhood

Driven by the discourse of well-being and the promise of preventing mental disorders, Aotearoa New Zealand has looked to educating SEL in schools as a way to support children and young people with attaining a 'healthy personality'. However, the focus on educating feelings is not new, as national legislation, guidelines, and curriculum have encouraged schools to focus on students' emotions for decades (see McLeay & Powell, 2022). Nevertheless, the drive for educating emotions has persisted with diverse 'ensembles' of students, parents, families, teachers, psy-professionals, academics, and state officials working together to promote SEL into the lives of all students (see, e.g., Borderless, 2020), including the very youngest people in early childhood. Although early childhood education is not compulsory in Aotearoa New Zealand, most young people have engaged with childhood centres or agencies (see Education Counts, 2022). Thus, the national early childhood curriculum, *Te Whāriki* (see Ministry of Education, 2017), serves as valuable documentary evidence of how emotions and well-being may be introduced into the lives of many young people across the country.

Te Whāriki: Promoting Well-Being in the Early Childhood Curriculum

The early childhood curriculum promotes well-being as a fundamental "right" that is to be "promoted and ... protected" (Ministry of Education, 2017, p. 19) for all young people in Aotearoa New Zealand. This championing of well-being as a 'right' sends a clear message to early childhood learning communities that well-being is something young people are entitled to from birth. Thus, various psychological theories, like holistic development (see Ministry of Education, 2017, p. 19) and positive psychology, have infiltrated educational pedagogy and become essential for supporting young people by promoting the belief that their well-being can be divided into various components (see Dalley et al., 2023) that can then be worked on, improved, and enhanced. The strategy of educating the individual well-being component (emotions) to enhance the whole person (also see Ministry of Education, 2023) has, therefore, directed schools to engage with and promote SEL approaches of helping young people with "managing themselves and expressing their feelings" as well as developing a "capacity for self-regulation [with their emotions]" (Ministry of Education, 2017, p. 27). However, achieving these curriculum goals cannot be left to chance and requires that early childhood educators draw upon intentional pedagogy (Ministry of Education, 2017), where they purposefully use "their knowledge, judgment, and expertise" (McLaughlin & Cherrington, 2018, p. 35) to strategically support children's emotional well-being. With the national curriculum firmly set on ensuring young people learn about well-being, a range of supplementary resources have proliferated (see, e.g., The Education Hub, 2024a, 2024c), including one particular government endorsed document that aims to help early childhood educators with supporting young people's emotional well-being: *He Māpuna Te Tamaiti* (Ministry of Education, 2019a).

He Māpuna Te Tamaiti: Emotional Competence for Emotional Well-Being

> Emotional competence includes understanding what we are feeling and why we are feeling it … It also includes responding to our feelings and expressing them appropriately – and, when the feelings are strong or difficult to manage, in ways that help us calm down and feel better. (Ministry of Education, 2019a, p. 35)

Translated as "each child is precious and unique" (Ministry of Education, 2019a, p. 7), *He Māpuna Te Tamaiti* promotes emotional competence (EC) as a primary way to support young people's 'right' to acquiring emotional well-being and a 'positive' personality (Ministry of Education, 2017). However, helping young people achieving EC in early childhood requires the use of various tactics and practices (explored in the following sub-section) and the use of one central governing element, mentioned below:

Relationships are at the heart of early learning. (Ministry of Education, 2019a, p. 23)

While young people experience a diverse range of relationships within their learning community, the teacher/student relation has been recognised as being significantly important for their learning (Ministry of Education, 2017). Subsequently, teachers are motivated to form positive relationships by fostering care, trust, and responsibility with young people (Ministry of Education, 2017). According to the Ministry of Education (2019a), these virtues may be developed through the consistent use of techniques such as appropriate gestures, vocal tone, or welcoming "[c]hildren … in the morning by name and in a manner that signals … positive expectations of a happy day ahead" (p. 25). By using techniques like these, young people learn they can "be themselves" (Ministry of Education, 2019a, p. 23), which opens them up to receiving guidance and being courageous in sharing who they are. This allows teachers to gain more insight about each student and also allows young people to learn more about themselves and their peers. Likewise, caring about the personalities, passions, "strengths, interests and areas for development" of students (Ministry of Education, 2019a, p. 23) supports the teacher-student relationship by enabling the teacher to tailor unique learning experiences and conversations that specifically scaffolds each child towards EC (Ministry of Education, 2019b).

The unique teacher/student relationship may also guide childhood educators to "manage and regulate" their own feelings (Ministry of Education, 2019a, p. 18), as they come to be highly regarded as experts in EC by the young people in their care. Accordingly, the Ministry of Education (2019a) and the New Zealand Teachers Council of Aotearoa New Zealand (2024) have encouraged teachers to attend professional development courses (see Virtual Lab School, 2023) and refer themselves to specialist services (like counsellors or psychologists) when needed, in attempts to maintain their well-being and emotions. Thus, EC drives educators and young people to monitor each other to ensuring everyone adheres to keeping emotions 'well'.

> The wellbeing of each child is interdependent with the wellbeing of their kaiako, parents and whānau. (Ministry of Education, 2017, p. 20)

While the teacher (Kaiako)/student relationship is important, the statement above also identifies that relations with parents and family (whānau) are also central for young people's well-being. Thus, the Ministry of Education (2017) guides early childhood educators to develop a caring teacher/parent relation where they monitor and, if needed, direct parents/families to access specialist well-being services and professionals. The virtues of care and respect are vital for the teacher/parent/family relation, as they stimulate parents and families to increasingly share information about their private and family lives, as well as encourage them to ask for the teacher's specialist expertise in EC and well-being (Ministry of Education, 2019a). *He Māpuna Te Tamaiti* also directs teachers to utilise the teacher/parent/family relation to gather insights about each family's understandings, knowledge, and rules with emotions (which are shaped by dominant discourses) for the co-construct of specific expectations for young people's EC (Ministry of Education, 2019a). Likewise, the directive to consult wider community members (Ministry of Education, 2017, 2019a) highlights how psy-professionals may (re) incorporate their dominant ideas and beliefs about feelings into community expectations of young people's emotional well-being. The goals of EC and well-being for young people are formulated through the strategic exercising of dominate beliefs between, within, and from teacher/parent/family/community that (re)produce and expand dominant psy-discourses—such as emotional well-being or troubled, deviant, youth (Cohen, 2016; McLeay et al., 2023). As a result, this network of relations guides how young people's emotions may be seen, thought, and spoken about (Rose, 1999). Although *He Māpuna Te Tamaiti* supports early childhood educators to strategise the introduction and promotion of emotional well-being within a learning community, the demand for daily tasks and activities that help young people develop social and emotional competences has led to the proliferation of practice-focused resources, such as the following national teacher resource (also see The Education Hub, 2024a & 2024b for more examples).

Intentional Teaching Practices for Educating Emotional Competence

The *Intentional Teaching Practice* cards (Ministry of Education, 2019b) illustrate a government endorsed resource for early childhood educators that provide them with specific "daily practices" and "professional learning conversations" (p. 2) to guide children towards EC. This resource encourages early childhood educators to utilise their unique expertise of each and every young person and knowledge of intentional pedagogy (Ministry of Education, 2019b) to exercise these practices in an efficient, flexible, creative, and focused manner.

Learning Observation

> "I noticed the way you … bounced back after you felt disappointed". (Ministry of Education, 2019b, p. 3)

The example provided above from the first teaching practice card, entitled *describing* (Ministry of Education, 2019b, p. 3), emphasises the importance of language for helping young people learn about their emotions. More specifically, by providing "verbal description[s]" (Ministry of Education, 2019b, p. 3), early childhood educators not only draw attention towards a variety of feelings (such as disappointment) and behaviours (like being resilient) but subtly guides young people to engage in the specific actions of looking and/or listening. Consequently, young people learn what information they will look and listen for on a daily basis and how to intentionally hone their attention and 'noticing'. The aim to make emotions and actions explicitly "clear to the child or other children" (Ministry of Education, 2019b, p. 3) illustrates how teachers may use the practice of describing to direct young people to look both inwards (at themselves) and outwards (at each other). As a result, young people learn essential movements for recognising their own (as well as others) emotions. Similarly, the gentle phrase "I noticed" (Ministry of Education, 2019b, p. 3) subtly draws attention to the caring gaze (Foucault, 1997) of the teacher, which stimulates young people to look towards them for their expert observations and guidance. Thus, the intentional teaching practice of *describing* supports young people with learning the movements of looking and listening to others (as well as themselves), which are essential skills for one of the components of the emotionally competent individual: *observation* (Ministry of Education, 2017, 2019a).

Talking About Emotions

> "I think you might be feeling frustrated, Rana, because you want to keep playing and it's tidy up time". (Ministry of Education, 2019b, p. 6)

The intentional teaching practice shown above of "providing vocabulary and information" (Ministry of Education, 2019b) illustrates how early childhood educators may work to "support children's understanding and ability to talk about feelings" (p. 6). Through associating emotions (such as frustration) to desires (you wanted to keep playing) and circumstance (it's tidy up time), early childhood educators support young people with learning how to 'connect details' together, which allows them to know how to speak about their emotions with increasing confidence. Similarly, other intentional teaching practices, such as *thinking aloud*—"talking aloud about what we are thinking, doing, or feeling and why" (Ministry of Education, 2019b, p. 12)—also work to boost young people's ability to vocalise and use specific information to communicate their emotions with others more effectively.

Self-examination is also an important part of the process of talking about emotions, as helping young people to reflect on their feelings supports them to speak about them. The following example illustrates how early childhood educators may deploy intentional teaching practices and techniques—such as *describing* and open-ended questions—to encourage young people to review and reflect on their feelings:

> "[Y]ou were pretty excited about the police dog visit today, but you stayed very quiet and calm. How did you do that?" (Ministry of Education, 2019a, p. 38)

This illustrates how young people can be directed to examine and review their feelings and actions, which are essential movements for discriminating feelings (Mayer & Salovey, 1993; Ministry of Education, 2019b). By using these intentional teaching practices, early childhood educators not only help young people understand what and how to speak about their feelings but encourage them to develop an awareness of who can, or even must, hear about their emotions, namely adults, experts, authorities, and other caring individuals. Consequently, young people learn the movements that are prime indicators of the emotionally competent individual: speaking about emotions and seeking helpful people to listen and care for their emotions (Ministry of Education, 2019a).

Practising Emotional Competence

> "OK now? Well, done for calming down. Would you like to talk a little about how you were feeling?" (Ministry of Education, 2019a, p. 44)

The learning conversation presented above demonstrates how early childhood educators may deliberately use the practice card of "providing opportunities to practise" (2019b, p. 21) to encourage young people to rehearse the movements learnt through the earlier intentional teaching practice cards. By creatively using additional teaching practice cards, for instance, like praise or encouragement, teachers may flexibly and 'positively' increase young people's opportunities and capacities to observe, talk about, manage, and regulate their emotions (Mayer & Salovey, 1993). As a result, young people learn how to demonstrate actions of well-being promoted in *He Māpuna te Tamaiti* and *Te Whāriki* (Ministry of Education, 2017, pp. 27 & 35) of self-control, self-calming, seeking help, or being resilient (Ministry of Education, 2017, 2019a) with increasing frequency and strength.

Becoming Emotional Competence 'experts'

> "I can give you a little reminder about calming down when you are cross. Soon you will know how to manage for yourself. And then you will be able to show others how we calm down". (Ministry of Education, 2019b, p. 7)

The intentional teaching practice of 'fading supports', illustrated above, provides an example of how early childhood educators encourage young people towards managing their emotions independently. Here, teachers promote young people's capabilities and subtly indicate that if they can learn to control their emotions, their intensive form of oversight will eventually no longer be required. Although gaining independence and taking responsibility over one's feelings demonstrates EC and well-being (Ministry of Education, 2019a, 2019b), the experience of a caring relationship, especially with authoritative figures like teachers, encourages young people to continually look to others for their views and guidance. While self-management is an important milestone of SEL and the national curriculum (Ministry of Education, 2007), the practice of *fading supports* guides individuals to take responsibility for the feelings and well-being of others (Ministry of Education, 2017). This emphasises a specific intentional teaching practice card that stimulates young people to form a unique relationship with their peers: *supporting tuakana-teina relationships*. The tuakana-teina relation requires students to take on a responsible caring role with others where "an older or more experienced child plays with and supports a younger or less experienced child" (Ministry of Education, 2019b, p. 15). Young people, who are competent in managing and regulating their emotions (Mayer & Salovey, 1993), may then recycle the teaching strategies used to support them with acquiring EC to help instruct the movements of their peers towards the co-constructed community expectations of emotional well-being (Ministry of Education, 2017, 2019a). Care and responsibility are central to this unique student-teacher/student relationship, as it drives the competent young person to support others with increasing intention, attention, and devotion.

Although the intentional teaching practice cards are one resource for supporting early childhood educators with teaching SEL and "do not represent an exhaustive or definitive list of practices" (Ministry of

Education, 2019b, p. 2), they do clearly illustrate the tactical guidance of movements that supports people in childhood learning communities to move in the direction of 'wellbeing'.

Discussing the Education of Emotions in Early Childhood

The Aotearoa New Zealand case study shows how early childhood social emotional learning uses a complex network of relationships to promote the dominant discourse of well-being, guiding young people towards a 'healthy identity'—defined as calmness, positivity, and happiness (Ministry of Education, 2019a). By framing well-being as a 'right', the Ministry of Education (2017, 2019a) highlights its strong historical alliance with psychological disciplines, urging learning communities to prioritise psychological concepts and expertise. This intricate network makes it increasingly difficult to avoid the influence of emotional education and resist psy-authority (Rose, 1999).

Within this intricate network of relations, the early childhood educator can be seen to assume the role of a 'shepherd' who, directed by the discourse of well-being promoted in the curriculum (Ministry of Education, 2017), takes increasing responsibility for the emotional competence of each and every young person in their care. Here, the directive to foster a 'positive relationship' of care and generosity (Ministry of Education, 2017) serves as a unique tactic that positively inspires young people to accept and follow the early childhood educator's guidance and instruction on how to move their minds, bodies, and souls (Binkley, 2014). Through this caring-responsible relationship, teachers become increasingly motivated to *surveil* and understand their students' emotional competence and, thus, seek out professional development opportunities (see Education Hub, 2024) to bolster their specialised knowledge of how to effectively monitor and help young people express their inner life. Although ensuring young people learn to control their feelings is important (Ministry of Education, 2019a), educating emotions encourages teachers to focus on helping their sheep learn the necessary actions that will allow them to move together in a shared direction (Foucault, 2009). The intentional teaching practices illustrate how early childhood educators, as shepherds, may utilise EC to guide the movements of children, directing them towards their salvation (well-being).

Early childhood educators support young people in learning two essential movements through educating EC: 'examination' and 'confession' (Binkley, 2014). Through the teaching practices of 'describing' (Ministry of Education, 2019b) young people begin to learn movements for examination, as they are guided to look intensely at themselves and to internalise an inspecting gaze (Foucault, 1980). Here, children and young people develop an individualising attitude and awareness of the self (Binkley, 2014) that supports them with observing and discriminating their own (as well as others) feelings (Salovey & Mayer, 1990). This development also illustrates the unique reciprocal relationship of the shepherd/flock, as the student gaze encourages teachers to monitor their own actions, thoughts, and feelings (Ministry of Education, 2019b). The movement of 'confession' may also be developed through intentional teaching practices, such as thinking aloud (Ministry of Education, 2019b) or learning conversation that encourages young people to talk about various ways to 'calm down' when feeling frustrated or excited (Ministry of Education, 2019a). These practices train young people to surveil themselves and confess deviations from the emotional norms set by the psy-focused learning community. By mastering self-examination and confession, young people provide pastoral authorities with insights into their inner world (Foucault, 1997; Hyman, 2014). As a result, the emotional competent young person may be instructed to seek out technologies and techniques for self-guidance that allow them to move their bodies towards the goal of happiness and well-being (Binkley, 2014; Rose, 1999).

Although teachers take up a shepherding role, practices like tuakana-teina (Ministry of Education, 2019b) show how the emotionally competent child may be incorporated as a shepherd who works to guide the movements of others towards emotional well-being. By using the intentional teaching practices they were guided by, young people adopt the shepherd/flock relationship and exercise pastoral power (Foucault, 1997) to inspire their peers to move towards emotional competence. This demonstrates how early childhood educators strategically work to positively produce young people as both 'sheep' and 'shepherds' who—through being encouraged to take increasing responsibility for the well-being and emotional competence of others (Ministry of Education, 2019a, 2019b)—come to independently care for the movements of their peers with increasing intention and intensity (Rose, 1999). Thus, educating young people in early childhood to be emotionally competent serves as a complex strategy that makes use of the unique relationships between teacher/student and student/student to guide a 'multiplicity of movements' (Foucault, 2009), which inspires individuals to learn how to move together in one direction and in unison (Foucault, 1997).

Ultimately, educating EC supports large populations to become increasingly manageable and controllable, as each and every individual learns specific movements (of examination and confessions) that support the endeavour to shape and *discipline* their minds, bodies, and souls (Rose, 1999). Pastoral power serves as a unique precursor to disciplinary power (Foucault, 1995), as the learning of movements authorises young people to become their own slave drivers (Scull, 1993) who work to positively produce a 'healthy identity' of well-being—happy, positive, optimistic. Thus, educating SEL and EC in early childhood epitomises Foucault's description of "the 'daemonic' coupling of 'city-game' and 'shepherd-game'" (Gordon, 1991, p. 8), where the management of populations is achieved through disciplinary, pastoral power and individualising practices of emotions that shape and direct how people see, think, speak, and respond (Rose, 1999) to young people's emotions.

Conclusion

By examining the rise of well-being, and the development and implementation of SEL in early childhood education, this chapter illustrates how educating emotions serve as a unique strategy that supports governmentality (Foucault, 1991), 'benevolently' utilising 'ensembles' of caring relations, strategies, tactics, and practices to shape the lives of young people (as well as other members of the learning community) towards certain ends (Foucault, 1991). The subtle way power-relations, discourses of well-being and happiness, as well as minute tactics to train movements are used to inspire children to become the task masters of their own (as well as others) emotions epitomises Foucault's notion of 'the conduct of conduct', which emphasises that modern government is "a form of activity aiming to shape, guide or affect the conduct of some person or persons" (Gordon, 1991, p. 2). While teaching EC in early childhood serves to strategically strengthen and reinforce the norms, knowledge, and authority of psy-professionals and industries, such as the happiness industry (Davies, 2015), a truly insidious ends of educating emotions may be seen in the way it benevolently coerces young people to (re)shape, utilise, and maximise their ability to choose well-being (happiness) regardless of circumstances. More specifically, the quest for a 'healthy identity' (the salvation of well-being) not only guides young people to effectively manage their emotions, but it also inspires them to become citizens who take increasing responsibility for using their individual liberty, autonomy, and rights to attain 'better' educational outcomes through managing their emotions. Thus, educating emotions serve as a strategy that supports young people to govern themselves (and others) towards becoming the ideal 'entrepreneurial' citizens (Ministry of Education, 2007), who 'happily' work on themselves to attain the character needed for the modern neoliberal workforce in Aotearoa New Zealand (also see Binkley, 2014; Cohen, 2016). The success of SEL to support the art of modern government has resulted in its permeation throughout society, from early childhood, primary, secondary, and tertiary education systems to business, health industries, and even family homes.

Consequently, there is a substantial amount of research and theoretical work that needs to be done in the field of educating emotions and well-being in schools. In particular, there is a need for critical scholarship that focuses on interrogating other contemporary psychological concepts used for educating emotions in schools (like 'emotional first aid') and for interdisciplinary studies (between sociology, psychological, education, and counselling, among others) that examine how power shapes the spaces, practices, and experiences of both the governors and the governed in school communities (see O'Malley, 2009). Additionally, education could benefit from further examining the way the discourse of well-being infiltrates their approaches and practices to supporting young people in schools. One way to achieve this is by producing more critical research that acknowledges the complexities and issues surrounding emotions, as well as their relations to specific social, cultural, and political contexts (Burkitt, 2018). By engaging in this theoretical work, the field of emotions (as well as well-being) could open up space for wider investigations, discussions, and debates about further cultural, collective, and social responses to supporting young people's emotions. Through these suggestions, new insights and opportunities for understanding how young people's emotions are constructed and shaped may arise (see also McLeay & Powell, 2022) that supports educational and psychological professionals in carefully considering and critiquing how their practices, services, theories, and knowledge serve young people in schools.

References

American Psychiatric Association. (2013). *Diagnostic and statistical manual of mental disorders* (5th ed.). American Psychiatric Association.

Bill & Melinda Gates Foundation. (2023). *Social and emotional learning.* U.S. Program. https://usprogram.gatesfoundation.org/what-we-do/k-12-education/social-and-emotional-learning

Binkley, S. (2014). *Happiness as enterprise: An essay on neoliberal life*. Suny Press.

Bond, T. N., & Lang, K. (2019). The Sad Truth about Happiness Scales. *Journal of Political Economy, 127*(4), 1629–1640. https://doi.org/10.1086/701679

Borderless. (2020, June 24). *MOE conference highlights* [Video]. YouTube. https://www.youtube.com/watch?v=iuGhcNgw_Vc

Burkitt, I. (2018). Decentring emotion regulation: From emotion regulation to relational emotion. *Emotion Review, 10*(2), 167–173. https://doi.org/10.1177/1754073917712441

Cabanas, E. (2016). Rekindling individualism, consuming emotions: Constructing "psytizens" in the age of happiness. *Culture and Psychology, 22*(3), 467–480. https://doi.org/10.1177/1354067x16655459

Cabanas, E., & Illouz, E. (2019). *Manufacturing happy citizens: How the science and industry of happiness control our lives*. Polity Press.

Carr, A. (2011). *Positive psychology: The science of happiness and human strengths* (2nd ed.). Routledge.

Child & Youth Wellbeing. (2019). *Overview: The framework*. https://childyouthwellbeing.govt.nz/sites/default/files/2019-08/strategy-on-a-page-child-youth-wellbeing-Sept-2019.pdf

Cohen, B. M. Z. (2016). *Psychiatric hegemony: A Marxist theory of mental illness*. Palgrave.

Collaborative for Academic, Social, and Emotional Learning. (2024a). *Fundamentals of SEL*. https://casel.org/fundamentals-of-sel/

Collaborative for Academic, Social, and Emotional Learning. (2024b). *Leading school-based teams through a process for systemic SEL implementation*. https://schoolguide.casel.org/how-it-works/

Corcoran, R., Cheung, A., Kim, E., & Xie, C. (2018). Effective Universal school-based social and emotional learning programs for improving academic achievement: A systematic review and meta-analysis of 50 years of research. *Educational Research Review, 25*, 56–72. https://doi.org/10.1016/j.edurev.2017.12.001

Dalley, J., Dixon, R., & Robertson, J. (2023). *Using Te Whare Tapa Whā for learning about wellbeing: activities for year 1-8 ākonga*. New Zealand Education Association.

Davies, W. (2015). *The happiness industry: How the government and big business sold us well-being*. Verso.

Duncan, G. (2019). Happiness and the new politicization of subjectivity. In N. Hill, S. Brinkman, & A. Petersen (Eds.), *Critical happiness studies* (1st ed., pp. 83–97). Routledge.

Education Counts. (2022). *Participation in early learning*. https://www.educationcounts.govt.nz/__data/assets/pdf_file/0009/219915/2022-Census-Early-Learning-Participation-Fact-Sheet.pdf

Education Hub. (2024). *Social Emotional Competence Course*. https://theeducationhub.org.nz/social-emotional-competence-in-ece-course/

Foucault, M. (1975). *The birth of the clinic: An archaeology of medical perception*. Vintage Books.

Foucault, M. (1980). *Power/knowledge: Selected interviews & other writings 1972–1977* (C. Gordon ed.). Pantheon.

Foucault, M. (1991). Governmentality. In G. Burchell, C. Gordon, & P. Miller (Eds.), *The Foucault effect: Studies in governmentality* (pp. 87–104). Harvester Wheatsheaf.

Foucault, M. (1995). *Discipline and punishment: The birth of the prison* (2nd ed.). Penguin. (Original work published in 1978).

Foucault, M. (1996). *Foucault live. Collected Interviews, 1961–1984* (S. Lotringer, Ed.). Semiotext(e).

Foucault, M. (1997). *Ethics: Subjectivity and truth. Essential works of Michel Foucault (Vol. 1)*. The New Press.

Foucault, M. (2009). *Security, territory, population: Lectures at the Collège de France, 1977–78*. Palgrave Macmillan. (Original work published in 1978).

Gard, M., & Pluim, C. (2014). *Schools and public health: Past, present, future*. Lexington Books.

Goleman, D. (1995). *Emotional intelligence*. Bloomsbury.

Gordon, C. (1991). Governmental rationality: An Introduction. In G. Burchell, C. Gordon, & P. Miller (Eds.), *The Foucault effect: Studies in governmentality* (pp. 1–52). The University of Chicago Press.

Gorton, D. (1873). *An essay on the principles of mental hygiene*. J.B. Lippincott & Company.

Holmes, M. (2021). Emotions in Aotearoa New Zealand: Reflexive emotionalisation in a colonised context. *International Sociology, 36*(4), 498–514. https://doi.org/10.1177/0268580921995301

Hyman, L. (2014). *Happiness: Understandings, narratives and discourses* (1st ed.). Palgrave.

Mayer, J., & Salovey, P. (1993). The intelligence of emotional intelligence. *Intelligence, 17*(4), 433–442. https://doi.org/10.1016/0160-2896(93)90010-3

McLaughlin, T., & Cherrington, S. (2018). Creating a rich curriculum through intentional teaching. *Early Childhood Folio, 22*(1), 33–38. https://doi.org/10.18296/ecf.0050

McLeay, R. (2025). Counselling power and knowledge in schools. In E. von Kardorff, M. Harbusch & D. Robin (Eds.), *Zur Gesellschaft der verletzten Seelen: Neue soziologische Perspektiven auf Psychiatrie, Gesellschaft und Subjekt*. Springer.

McLeay, R., & Powell, D. (2022). Governing emotions in school. In M. Harbusch (Ed.), *Troubled persons industries: The expansion of psychiatric categories beyond psychiatry* (pp. 49–76). Palgrave Macmillan. https://doi.org/10.1007/978-3-030-83745-7_3

McLeay, R., Powell, D., & Cohen, B. M. Z. (2023). Young people's voices on emotions: A narrative inquiry. *International Journal of Qualitative Studies in Education, 37*(2), 342–357. https://doi.org/10.1080/09518398.2023.2233945

McLeod, J., & Wright, K. (2016). What does wellbeing do? An approach to defamiliarize keywords in youth studies. *Journal of Youth Studies, 19*(6), 776–792. https://doi.org/10.1080/13676261.2015.1112887

Mental Health & Wellbeing Commission. (2023). *Opinion piece: Our priorities for New Zealand's First Minister of Mental Health.* https://www.mhwc.govt.nz/news-and-resources/our-priorities-for-the-incoming-government/

Mental Health Foundation. (2023). *Statistics on mental health in schools in New Zealand.* https://mentalhealth.org.nz/statistics-on-mental-health-in-schools-in-new-zealand

Ministry of Education. (2007). *The New Zealand curriculum*. Learning Media.

Ministry of Education. (2017). *Te whāriki: Early childhood curriculum*. Ministry of Education.

Ministry of Education. (2019a). *He Māpuna Te Tamaiti: Supporting social and emotional competence in early learning*. Cognition Education for the New Zealand Ministry of Education.

Ministry of Education. (2019b). *Intentional teaching practices: He Mapuna te Tamaiti, Supporting social and emotional competence in early learning*. Ministry of Education.

Ministry of Education. (2023). *Understanding social and emotional learning. Health and physical education.* https://hpe.tki.org.nz/professional-learning-support/understanding-social-and-emotional-learning/

O'Malley, P. (2009). Governmentality and risk. In J. O. Zinn (Ed.), *Social theories of risk and uncertainty: An introduction* (pp. 52–75). Blackwell.

OECD. (2023, March 16). *Schools as hubs for social and emotional learning – Are schools and teachers ready? Education spotlight*. OECD.

Park, N., Peterson, C., & Seligman, M. E. P. (2004). Strengths of character and well-being. *Journal of Social and Clinical Psychology, 23*(5), 603–619. https://doi.org/10.1521/jscp.23.5.603.50748

Pauwels, B. G. (2015). The uneasy – And necessary – Role of the negative in positive psychology. In S. Joseph (Ed.), *Positive psychology in practice: Promoting human flourishing in work, health, education, and everyday life* (pp. 807–822). John Wiley & Sons.

Rose, N. (1999). *Governing the soul: The shaping of the private self* (2nd ed.). Free Association Books.

Ryan, J. (2024). *Controller & auditor-general: Meeting the mental health needs of young New Zealanders*. https://oag.parliament.nz/2024/youth-mental-health/docs/youth-mental-health.pdf

Salovey, P., & Mayer, J. (1990). Emotional intelligence. *Imagination, Cognition and Personality, 9*(3), 185–211. https://doi.org/10.2190/dugg-p24e-52wk-6cdg

Scull, A. (1993). *The most solitary of afflictions: Madness and society in Britain, 1700–1900.* Yale University Press.

Seligman, M. (2002). *Authentic happiness.* Free Press.

Shadé, Z. [@shadezahrai]. (2022, April 26). *Instantly Boost Emotional Intelligence | #shorts* [Video]. YouTube. https://www.youtube.com/shorts/n1dPik4LzIE

Teachers Council of Aotearoa New Zealand. (2024). *Teacher wellbeing.* https://teachingcouncil.nz/resource-centre/teacher-wellbeing/

The Education Hub. (2024a). *Social and emotional competence in early childhood education.* https://theeducationhub.org.nz/category/ece-resources/social-and-emotional-competence-in-ece/

The Education Hub. (2024b). *Social and Emotional Learning* (SEL). https://theeducationhub.org.nz/category/school-resources/social-emotional-learning-sel/

The Education Hub. (2024c). *Supporting children's emotional learning in early childhood.* https://theeducationhub.org.nz/supporting-childrens-emotional-learning-in-early-childhood/

Virtual Lab School. (2023). *Social emotional learning for teachers: An introduction.* https://www.virtuallabschool.org/focused-topics/social-emotional-learning-for-teachers/lesson-1

Walsh, C. (2024, January 15). *Best Social Skills Apps for Kids.* Educational App Store: Supporting the digital wellbeing of children. https://www.educationalappstore.com/best-apps/best-social-skills-apps-for-children-and-teens

Watt Smith, T. (2015). *The book of human emotions.* Profile Books.

Wright, K. (2014). Student wellbeing and the therapeutic turn in education. *Australian Educational and Developmental Psychologist, 31*(2), 141–152. https://doi.org/10.1017/edp.2014.14

Part II

Empowering C/S/X in Research and Praxis

5

Co-producing Histories of Australian Community Mental Health, 1970–1990

The Re;Minding Histories Research Group

Dedicated to the memory of Ruah Potaris Grace

Introduction

Over the past few decades, researchers and policymakers developing or modifying mental health programmes and services have invited individuals with a lived experience of extreme mental distress to participate in research and planning projects. Because of their history of experiencing

Members of the *Re;Minding Histories* Research group include Samantha Baker, Roslyn Burge, Carla Gaskin Charles, Catharine Coleborne, Robyn Dunlop, Lewis Gould-Fensom, Ruah Potaris Grace, Anthony Harris, Tina Kenny, Hans Pols, Paul Rhodes, Ian Shoebridge, Gemma Lucy Smart, Georgia Valis, Eloise Watts, and Asha Zappa. Contributors to this chapter are Samantha Baker, Roslyn Burge, Catharine Coleborne, Anthony Harris, Hans Pols, Paul Rhodes, Gemma Lucy Smart, Georgia Valis, Eloise Watts, and Asha Zappa.

The Re;Minding Histories Research Group (✉)
The University of Newcastle, Newcastle, NSW, Australia
e-mail: catharine.coleborne@newcastle.edu.au

N. A. Hendry, E. Karageorgos (eds.), *Critical Mental Health in Australia and Aotearoa New Zealand*, The Politics of Mental Health and Illness,
https://doi.org/10.1007/978-3-031-82861-4_5

mental distress and accessing mental health services, peer researchers can make meaningful and significant contributions to mental health research. Instead of being objects of investigation who are, at times, consulted, these peer researchers have become valued members of research teams. Research projects that invite the active and full participation of various stakeholders as co-investigators tend to follow co-production methodologies, which value and recognise the unique knowledge and insights of all members of a research team. For consumers of mental health care, participating in co-production research signifies a significant shift from past experiences, as the perspectives of those considered "mad" have been "historically and systemically undermined and entirely dismissed from consideration" (Joseph, 2019, p. 2). In our research project on the history of community mental health in Australia from 1970 to 1990, we developed methodologies of co-production for historical research. We invited several lived experience researchers to join the research team along with a psychiatrist, a clinical psychologist, two historians of mental illness and psychiatry, a public historian, and several honours and postgraduate students with backgrounds in history, history and philosophy of science, medicine, and psychology. In this chapter, we provide an overview of the activities of the research group and reflections on our approach. This chapter is a partial report on how co-production methodologies can be applied in historical investigations of the recent past of mental illness (see also Ellis & Coleborne, 2022).

Background

Psychiatry, initially named alienism or mad doctoring, originated after physicians began working for mental hospitals, originally known as "asylums," to keep an eye on the health of those confined there. Being able to observe and investigate large numbers of "mad" people, they soon made "madness" an object of medical inquiry. According to Michel Foucault, the (French) Enlightenment started with the institutional confinement of those whose behaviour transgressed the standards of rationality. As he famously stated: "modern man no longer communicates with the madman. … The language of psychiatry, which is a monologue by reason about madness, could only have come into existence in such a silence" (Foucault, 2006 [1961], p. xxviii). Inspired by Goffman's (1961) trenchant critique of mental hospitals as total institutions like armies and concentration camps, several historians painted bleak pictures of asylums in the nineteenth century, characterising them as places of confinement, seclusion, neglect, and, at times, abuse and cruelty (see, e.g., Rothman, 1971; Scull, 1979; Porter, 1987).

In the 1970s, several historians started conducting research in social history in reaction to their more traditional colleagues who had focused on leaders of state, generals, and upper-class citizens; thus far, the lives of common labourers, the poor, the disadvantaged, and the downtrodden had escaped historical attention. Inspired by similar motives, feminist historians started investigating women's history by focusing on the domestic sphere. Research in social history is inspired by the civil rights and anti-authoritarian movements of the 1960s, which included the anti-psychiatry and the mental health consumer rights movements. In the 1980s, medical historian Roy Porter (1985) admonished his colleagues to adopt this new approach when researching the history of medicine, which he claimed should be conducted "from below" or from the patient's perspective. In response, social historians of psychiatry began to focus on recovering and reconstructing the voices of the mad, judiciously analysing patient records and other archival sources. They generally characterised early mental hospitals as oppressive places but have also highlighted the frequent interactions between superintendents, physicians, the "mad," and their families (Coleborne, 2006; Edington, 2019; Tomes, 1994). *Pace* Foucault, there was no total rupture and the dialogue between these actors in mental hospitals continued, although large power differences shaped these interactions.

Historians face several challenges when they attempted to make the voices of former mental hospital inmates audible. Physicians have published books and articles while mental hospitals published annual reports, but those confined within them left fewer records. Their perceptions, opinions, reactions are therefore only uncovered by reading those sources "against the grain"—a method Carlo Ginzburg (1980) developed to reconstruct the beliefs of individuals who were tortured by the Spanish Inquisition. Reconstructing patient voices through the institutions that confined and controlled them signals a disruption of the power structures of both hospitals *and* archives (Coleborne, 2018, 2021).

Historians realise such disruption by reconstructing the details of patient lives or by collaborating with genealogists, or by using genealogical methods (Watts, 2023). Record linkage serves as a powerful tool to demonstrate how individual lives surface in a range of institutional archives, making it possible to investigate people whose lives were hidden from view because of poverty, illness, or neglect (McCarthy et al., 2017). Investigating material culture has become an important tool in investigating the lives of people confined to asylums. Physical and oral traces, in addition to memories of their presence and experiences, can provide significant insights into the lived experience of mental illness (Bronwyn, 2010). Museum exhibitions have provided additional ways of presenting alternative histories of mental illness and institutional confinement (Coleborne & MacKinnon, 2011).

Social historians have piloted the use of interviews in oral history because it promised to provide access to new voices, information, experiences, and subjectivities. Controversy about the reliability and accuracy of the recollections thus recorded resulted in new theorisations of memory, collective memory, and composure (Perks & Thomson, 2016 [1998], pp. 4–5). For historians of psychiatry and mental health, the value of oral history lies in accessing recollections of institutional experiences. Using oral history has changed how historians conceive of authorship. In the discipline of history, sole-authored books and articles are still common; historians have theorised narrative authorship as a legitimate mode of writing about the past (Green & Troup, 1999; Munslow, 2012). Foucault (1998 [1969]) has argued that the name that appears as the "author" of a text does not refer to a person but is a construction of the "legal and institutional systems that circumscribe, determine, and articulate the realm of discourses" (p. 309). In the sciences, the assignment of authorship and, thereby, intellectual property and credit, has been analysed extensively (see, e.g., Biagioli & Galison, 2013). Scholars in the humanities continue to adhere to the ideal of single authorship as evidence of originality, authority, and responsibility. According to Adema (2021), these notions are based on humanist and romantic notion of the individual author as genius.

As the discipline has changed, especially in light of new social history methods, some historians have experimented with new models of collaboration and authorship. This is especially true for social histories of communities using oral historical approaches and public histories that require an engagement with wider audiences (Clark et al., 2018). Activist researchers have used co-authorship as a radical political act to explicitly include outside "voices" and meanings, arguing that doing so provides a powerful storytelling tool, especially in oral and digital history projects that seek the involvement of community members as research participants (Atlas, 2020). Influential historian Allesandro Portelli (1991), best known for his oral history work, suggests that interviews involve "two subjects who recognize each other as subjects, and therefore separate, and seek to build their equality upon their difference in order to work together" (p. 43). Arguing that oral sources "are always the result of a relationship, a common project," Portelli and his colleagues were aware that their emphasis on the importance of oral sources—as opposed to the "facts" of the story—unsettled the history establishment (Portelli, 1981, p. 103).

We also aim to unsettle the authorial voice by involving and including different voices and recollections, starting with the words of two historians, and then involving the team as a collective. Historian of psychiatry and member of the research team Catharine Coleborne (Cathy) has conducted extensive research using patient records, in one case to investigate the challenging and often difficult experience of migration from England to Australia and Aotearoa New Zealand (Coleborne, 2015). As individuals were committed to institutions most often by their families, their records provide extensive information on affected individuals and family dynamics. Hans Pols has previously investigated the history of the American and international mental hygiene movement (Pols, 2001, 2010), which was primarily a form of intellectual history. Their commentaries, which touch on the historiographical field, are designed to be useful to readers of this chapter in their acknowledgement of the shifts they have made as historians whose work led them to a new collaboration and form of research in an interdisciplinary and cross-institutional team, as examined in this chapter. Their words also promote the shift from "I" to "we" in our research.

Cathy

Having examined the role of families and their interactions with institutions in the nineteenth century, my scholarship evolved further when I joined a community-based project in 2003 involving a collaborative oral history project in Aotearoa New Zealand to capture the stories around the closure of Tokanui Hospital in 1998. My archival work acquired a different significance and meaning (Coleborne & The Waikato Mental Health Research Group, 2012). I observed how visitors to the museum exhibition about Tokanui at Te Awamutu Museum engaged with the displays and wanted to bring this history alive by liberating it from the archival records. They frequently remarked on the carceral aspect of the institution after seeing a heavy old ward key that was part of the exhibition. Even though the research team conducted almost fifty oral history interviews, it proved difficult to interview former service users. I was interacting instead with mental health service providers at a time of political change and dissent around psychiatric treatment, confinement, and allegations of abuse in Aotearoa New Zealand's mental hospitals. At that time, the Confidential Forum for Former In-Patients of Psychiatric Hospitals (2004–2006) was informing new understandings about mental health services. I recall feeling very uncomfortable about requiring those interviewees who had been former "patients" to sign non-disclosure agreements, which was off-putting to potential interviewees and former staff who wanted to discuss their experiences with the people they had cared for. We only interviewed two people who had been hospitalised at Tokanui. I visited them at their homes, recorded the interviews and transcribed them, and used the transcripts in my research and writing after asking the interviewees' permission.

Later, I was inspired by the work of Mary O'Hagan, a peer worker, mental health advocate, and former Aotearoa New Zealand mental health commissioner. After interviewing Mary, and reading more about "mad studies," I gained new insights about how historical research should be conceived and conducted; it became clear to me that we needed more radical approaches to include consumer voices (see Beresford & Russo, 2022; Gorman & LeFrançois, 2018; Menzies et al., 2013). One way to achieve this is through organising "witness seminars" (Nicholls, 2022; Tansey, 2007) in which participants are asked to reflect on their memories and past experiences. In my reflections on why historians should talk about madness, I outline O'Hagan's research agenda to tell the stories of hurt, harm, and hope in mental health, which has shaped my role in the research team (Coleborne, 2020, pp. 53–61, 65–58).

Hans

My earlier research did not focus on consumers of mental health care but on the ideas of psychiatrists and mental health professionals in the Western world. I later became interested in the history of psychiatry in the Dutch East Indies and analysed how Dutch colonial psychiatrists viewed the "native mind," which they characterised as primitive, emotional, impulsive, and inherently deceptive (Pols, 2007). The most interesting part of this research was the response by Indonesian physicians to these ideas, which they artfully deconstructed. While in Indonesia, I became increasingly interested in the recent history, current challenges, and future possibilities of psychiatry and mental health services there. With several colleagues, I started conducting research on this topic, using approaches I had thus far been unfamiliar with: we conducted interviews, held workshops, and visited sites. I established contact with several consumer support groups and was impressed by their ideas, activities, originality, and vibrancy. I attended meetings, training sessions, public activities, and had conversations with members. For an edited book that came out of this research project, I invited several members of these groups to write about their experiences, their sources of support, and the history of their support groups (Pols, 2020; Pols et al., 2019).

Community Mental Health in Australia

In Australia, idealistic psychiatrists, psychologists, social workers, mental health nurses, and occupational therapists started community mental health initiatives in the 1970s, more than a decade after deinstitutionalisation had commenced. In these initiatives, they frequently associated with individuals who had been discharged from mental hospitals; many pioneering small civic associations (currently named Non-Government Organisations or NGOs) provided innovative services to individuals who had been discharged from mental hospitals and gave them meaningful roles in running these associations, thereby developing the concept of peer workers. In Australia, the Consumer/Survivor/Ex-patient (C/S/X) and community mental health movements were inextricably linked (Dunlop & Pols, 2022). Investigating these early initiatives meant that traditional academic research was not appropriate for this project; the academics who became part of the research team realised that examining the history of Australian community mental health without the involvement of peer researchers would be highly problematic (Coleborne, 2020). Initiating a co-produced research project focusing on listening, interviewing, and storytelling was a decision we made collectively. When we started, we hoped that our research would enable and foster new and emerging narratives on the experience of extreme mental distress, mental health, and mental health services that will reshape future practices (for outputs from this research project other than those cited see a special issue of *Health and History* entitled (Re)Writing Care (Smart & Zappa, 2022) or consult the website of the research team (Re;Minding Histories Research Group, 2024)).

Co-production

Over the past twenty years, there has been a notable increase in research projects following the principles of co-production, inspired by notions of democracy, social engagement, social justice, and inclusion. Instead of consulting stakeholders and relevant parties, interviewing them, or holding focus groups, co-production research invites all relevant parties to become involved in the research process as co-investigators alongside academics and policymakers. Co-production research is used to address a broad range of issues at all levels of government as well as by health institutions to develop or revise policies and practices (Elroy, 2017).

Sheila Jasanoff (2004), a leading scholar in Science and Technology studies, has advocated co-production in developing policies in areas such as artificial reproduction, genetic engineering, climate change, and other areas where both scientific research and broader social concerns are central.

In research focusing on developing or changing mental health services or programmes providing various forms of social support, co-production practices are increasingly common. In a comprehensive review, Brett et al. (2014) concluded that co-produced research tends to be of higher quality and relevance. Co-produced mental health and social support services report better outcomes and higher levels of satisfaction among consumers of mental health care. Co-production "improves the relevance and resonance of research priorities, outcomes and [raises] the quality of research interpretation and knowledge translation" (Bellingham et al., 2022, p. 4). Involving consumers of mental health services in mental health research projects is revolutionary, as their views have traditionally been dismissed as biased, irrational, misguided, which has devalued any insights they could offer. Yet, as Roper et al. (2018, p. 4) state: "consumers have developed ways of knowing, theorising and thinking about their experiences that constitutes a unique discipline in the field of mental health known in Australia as consumer perspective." Inspired by co-production research projects in mental health, we decided to adapt co-production methodologies for our research on the history of Australian community mental health initiatives. We started with two excellent Australian guides on co-production in mental health (Bellingham et al., 2022; Bellingham et al., 2023; Roper et al., 2018) and adapted them for research in the humanities.

Co-production research follows several principles. First, co-production research projects deliberately and actively seek the involvement of the individuals and groups who have traditionally been the object or envisaged beneficiaries of research to become valued and equal participants in the research project. This respects one of the central demands made by the disability movement: "Nothing about us without us." The involvement of lived experience experts aims to counter the epistemic injustice such individuals have experienced in the past and are often still experiencing (Fricker, 2007; LeBlanc & Kinsella, 2016). Second, lived experience researchers have to be involved in *all* stages, including "planning, designing, conducting, and disseminating research" (Bellingham et al., 2022, p. 3). They should be given leadership roles to counteract past marginalisation and disenfranchisement. Ideally, they constitute at least *half* of the research team. Third, research teams should focus on gaining, building, and maintaining trust among all participants by "positioning consumers as leaders and creating safe and effective environments where consumer knowledge and expertise is valued and prioritised" (Roper et al., 2018, p. 4). This requires that power differences related to status, qualifications, and institutional position should be acknowledged, negotiated, and openly discussed. Fourth, peer researchers, who are often precariously employed, should be remunerated for their efforts, as other researchers hold continuing positions which provide them with a reliable source of income. Engaging in co-production research can be challenging and all team members should be willing to learn. As Bellingham et al. (2022) state: "Authentically engaging with a diversity of perspectives can lead to uncomfortable conversations."

Reflections: The Research Team

Our research project encountered its first challenge when we applied for funding from the Australian Research Council (ARC). The format, rules, and requirements of the ARC did not allow us to appoint lived experience researchers as equals to the chief investigators (CIs) and partner investigators (PIs). Hans reflects: "We had to be creative. Most consumer advocates do not have academic jobs and are generally only employed part-time, so they cannot be listed as Chief Investigators (CIs) or Partner Investigators (PIs). The only way we could provide them with an income was to hire them as Research Assistants (RAs), but then they had to occupy subordinate positions in the research hierarchy. RAs are not mentioned by name in grant applications and are not expected to be involved in grant writing. They are expected to work under the guidance of the CIs and PIs and do not own any intellectual property in their work or the research outputs of the research team. Early on, we decided to ignore these conventions. If our publications have twenty authors, that merely reflects the nature of our work. In that way, we can credit everybody." These practical considerations became a constant topic of conversation and innovation, providing opportunities to address the field. Another issue was providing payment to the consumers we interviewed; as most are precariously employed, we wanted to reimburse them for their time. Our universities prefer to provide gift cards, which we thought was inconsistent with the principles of our research project. Human research ethics committees prefer not to offer "incentives" that could pressure potential interviewees to participate. Despite this, we found a respectful way to provide fair remuneration to consumers who were interviewed.

We commenced the project with the involvement of our researchers on 25 April 2020. We hosted a day-long workshop on co-production led by Jo River and Brett Bellingham (University of Technology Sydney) who both have extensive experience in working with consumers in co-production research projects and are among the co-authors of the *Co-Production Kickstarter* (Bellingham et al., 2022) and the *Co-Design Kickstarter* (Bellingham et al., 2023). At that time, we were in lockdown because of COVID-19; the workshop was therefore held online via Zoom, which is not an ideal medium for workshops where people are meeting each other for the first time. During the workshop we agreed on several ground rules, which included openness, transparency, and mutual respect as well as a commitment to co-production, and formulated a group agreement (see Appendix I). We also agreed that at least half of all interviews would involve consumers, primarily focusing on the pioneers of the consumer movement, because consumers had been left out in earlier research. To guarantee the inclusion of a consumer perspective at each interview, we agreed that at least one lived experience researcher would be present at each interview. This changed the interview process in many positive ways. At some team meetings, we hosted other visitors. Kate Davies, then at the University of Newcastle, explored how to create "plain language" resources. Cathy recalls that this "was a profoundly interesting session that reminded all of us about the role of language in mental health research, and about the importance of accessibility as we confront and engage with all forms of disability in our communities." We later hosted Anto Sg, an Indonesian mental health activist, who shared his experiences of ill-treatment in Indonesia's mental health system and told us about various consumer initiatives in that country.

Following the workshop we started meeting every fortnight. During the first team meetings, the goals of the research project evolved and expanded, which was primarily driven by the lived experience researchers. One key change was including the history of the Australian mental health consumers' rights movement because former inhabitants of mental hospitals had played significant roles in the early community mental health scene. Their stories could not be told separately as they were intertwined. It was already clear that the Australian consumer movement had not received much scholarly or public attention; many current consumers are unaware of the accomplishments of the movement's pioneers. The research consequently became much more interesting and captivating. As Holly, one of the lived experience researchers on our team, remembers: "Exploring the various theories of change that C/S/X movement leaders deployed might assist mad pride activists in better navigating the movement's fraught internal politics, including lateral violence, learning from hard-won lessons of the past, and celebrating our elders on whose shoulder we stand." This goal was added to the planned outcomes of the project and became a shaping idea for us all.

During the meetings of the research team, we discussed how the results of our research project would be disseminated, for which audiences they would be written, and who would be able to access them. Articles in academic journals generally address academic audiences and are often behind paywalls, only accessible by people associated with universities. The lived experience researchers insisted on making our research findings accessible to everybody, in particular to consumers of mental health services, peer workers, mental health activists, and others with an interest in mental health and mental health services. Eloise, an honours student and research assistant, remembers that the lived experience researchers "were more interested in how our research would benefit people who are currently experiencing mental distress. They wanted to feed the research outputs back into the community and definitely had more of a grassroots focus." Samantha, an honours student and research assistant, agrees: the peer workers "wanted someone sectioned in a psych ward to be able to pick up on our history." We organised several witness seminars with the consumer community mental health movement to share their memories. Holly enjoyed facilitating these events: They were "a highlight for me personally because I got to be involved in 'research dissemination' that makes knowledge freely available to folks outside the academy and to use my facilitation skills which are much more adept than my academic writing skills." We set up a website featuring short informative pieces and art made by consumers. We organised a panel at The Mental Health Services (TheMHS) conference held in Sydney in October 2022; presentations by the consumer researchers were the centrepiece. Asha, one of the lived experience researchers, remembered it well: "I loved that. I think I do remember getting a lot of positive feedback after that. And it was on my birthday."

Academic psychiatrist Anthony had ample experience in working with consumers of mental health services and carers through his involvement with a mental health NGO before joining the research team. He reflects that he had not been involved in a project "that explicitly included consumers in the decision-making process." He continues: "Given the aims of the project to examine community psychiatry from all perspectives, the inclusion of lived experience researchers made complete sense and opened the door to perspectives that had not been included in many of my projects to that point." According to Anthony, adopting co-production methods "flattened the decision-making structure." To work with a team so deliberately diverse was exciting and at times captivating—there were so many different, illuminating perspectives and voices during our meetings. Georgia, an honours student and research assistant, reflects: "The project and meetings felt exciting because all of these different voices came through in our discussions, analyses, and outputs. These contributions offered different, sometimes disparate points of view, which were not necessarily resolved, but always valued and engaged with respectfully." Having many opinions, approaches, and perspectives made the meetings of the research team much more engaging and rewarding. According to Eloise, the consumer researchers "got this unique experience" that "added a lot of vibrancy to the team meetings." Georgia reflects: "The shared commitment to the goals of the project and the consensus to base our methodology on the 'lived experience' approach engendered a level of respect and encouragement amongst team members that felt productive and refreshing." She valued her experience in the research team highly: "I was silent at first, fresh from a degree in medical science and acutely aware I had none of the group's vast experience. Far from remaining an observer, all the researchers included me as one of the team, so I learned very quickly and found the right questions to ask." According to Eloise, the "consumers definitely had a grounding presence in the project. I think they kept us accountable in certain ways. And I think they unlock a level of humanness and openness and heart that I haven't seen in other groups that I've worked in since."

To reverse a long history of epistemic injustice and exclusion, guidelines for co-production recommend giving lived experience researchers leading positions in the research team. Georgia appreciated how "the lived-experience researchers ran with the opportunity to lead and shape the research project, teaching by example, and most notably establishing the significance of the consumer movement within the scope of the project, during the initial design process." We soon started rotating the position of chairing the meetings and found that Holly had a real talent in doing this. Asha ("I LOVE a good spreadsheet!") took up the role of making a comprehensive spreadsheet with the names of all past and future interviewees, making sure that various professions (psychiatrists, psychologists, occupational therapists, nurses, and others) were represented, as well as various demographic variables (age, gender, sexual orientation, geographic location).

The inclusion of lived experience researchers (Holly, Tina, and Ruah; Ian and Asha followed later) had an incredible impact on the research team. A key contribution they made was starting all meetings with "checking in," which involved inviting all researchers present to share a meaningful experience they had over the last week or to simply share how they were feeling that day. These check-ins created a space for all of us to be as open and candid as we wanted. As a result, we treated each other with empathy and started to feel comfortable when vulnerable. All meetings ended with a check-out, to make sure that the meeting had gone well and that there was no "unfinished business" to attend to. Hans reflects: "Group dynamics are essential; we needed to ensure that everybody felt like a participant and that nobody felt excluded. The knowledge produced depended on it. … Checking in at the start makes sure that everybody feels involved and respected." Georgia agrees: "Though initially uncertain about the process of book-ending team meetings with individual check-ins and -outs, the benefits of dedicating time for everyone present to share with the team an update of 'where they were at' gave context to their presence on that day, and also presented the opportunity to set or clarify expectations. The [check-ins] were genuinely helpful and fostered group cohesion."

The inclusion of consumer researchers not only made the project more inclusive of new perspectives and different forms of knowledge, it also changed the nature of our research meetings. The lived experience researchers taught the "conventional" researchers to respect all contributions and to allow all participants to freely express both opinions and emotions. Roslyn Burge, a public historian with a strong interest in mental health, particularly around the history of Callan Park mental hospital, recalls: "Fortnightly Zoom meetings on Friday afternoons … could have been a swift exercise in abandonment. Instead, these meetings were energising, challenging, and rewarding as everyone attending had an opportunity to contribute and comment on their experience of recent interviews. … A democratic space of review and co-production opened, and an exemplar of how to manage a significant nation-wide project by co-production during COVID—or any time." According to her, the lived experience researchers brought "vigour" to the discussions and "liberated the conversation with humour and irreverence." Cathy noticed her own attention to the driving questions and the form of the engagement in ideas changed her perspective on research teams and their conduct. "It was liberating and rewarding." In many ways, the form of research is allied to public histories of mental illness in museums (see Ellis & Coleborne, 2022).

The peer workers regularly expressed "Big Feelings," which made the other participants increasingly comfortable to do the same. Paul, an academic clinical psychologist, highlighted "witnessing the self-disclosure of our lived experience colleagues, compared with the taboos of the field of clinical psychology, [and the ensuing] discovery of new languages that can explain distress without resorting to diagnostic nosology." Eloise was "surprised with how candid the interviewees were and how open they were in telling their life stories, to open up and discuss very vulnerable parts of their lives. That was very much driven by the consumers in the group. They facilitated the interviews quite beautifully because of their own candidness and openness. … They shared from the heart quite a lot and grounded everything we did in a very authentic place." Gemma, an advanced postgraduate student, states: "I was and am in constant admiration of the ability of the peer workers in our group to handle 'Big Feelings' and it has become something I now aspire to in my own career, and life." Holly reflects: "I think we all did an incredible job at the relational aspects of the work of co-production and fusing lived and learned expertise." We were "moving at the speed of trust."

Research following the principles of co-production aims to reduce power differences in the research process or, at the very least, to discuss these power differences and their effects on the research openly. Asha remembered that "most of the time, lived experience voices were really central and really valued. The approaches that we took to the research were valued as well. Especially with the check-ins created a space where everybody, not just the lived experience researchers, took some time to reflect on their feelings and take responsibility for their feelings. I think that allowed us to have pretty robust discussions." Creating a space of empathy and respect not only allowed the lived experience researchers and some of our interviewees to share deeply painful memories of the time they spent in mental hospitals and psychiatric wards (at times against their will), it also allowed other team members to share relevant experiences they would otherwise have kept to themselves. It served as a constant reminder of the challenges individuals experiencing mental distress face, not only because of their mental state but even more because of the ways in which Australia's mental health services responded to them.

The interviews became the heart of our research project; we conducted more than one hundred (for our interview guide see Appendix II). Newspaper articles, official publications, annual reports, and various forms of "grey" literature all contain important information, but we were convinced that the memories of all individuals who were involved in early initiatives in community mental health would provide essential insights often absent in these sources. Many interviews were conducted online using Zoom, initially because of the lockdowns, but later because it enabled us to increase the geographic reach of the project. Because of the central role of interviewing, Ruah, a lived experience researcher, named our research team a "listening tribe." During the interviews, co-production approaches were evident. As Asha recalls, we approached "the interviews from a compassionate place, with a lot of kindness and curiosity. We were able to hold space in a way that allowed other people to have the freedom to think about things and be curious about themselves. It was holding the space, and I think we did that really well." They thought that the "lived experience researchers bring a different energy and dynamic to the interviews."

Cathy, reflecting on the interviews she co-conducted, remembers: "In my interviews with Sandy Jeffs (a poet and person with lived experience of institutions; see Jeffs & Leggatt, 2020; Jeffs, 2022), I worked alongside Holly. This was my first experience of interviewing as a 'team', and it was a much richer and rewarding experience. I felt supported as a historian to venture into difficult aspects of the discussion, safe in the knowledge that we had created a space for dialogue. Holly and I both loved Sandy's concept of her support during periods of illness and recovery: she called it 'Team Sandy'. There was a lovely sense that we had learned together." Cathy also remembers her interview with early mental health advocate Sandy Watson: "She was able to take us back to sites in Sydney and her early activities to organise consumers of mental health care, including through cartoons she created" (see also Pols & Grace, 2022). As Roslyn, who has extensive experience in conducting oral history interviews, recalls: "Initially I was surprised by the informality and unstructured way in which some interviews commenced. It very quickly became evident, time and again and in multiple interviews, that interviewees given such freedom quickly became comfortable. … Given such agency, not one interviewee hesitated, and the interviews followed the energy and direction of the interviewee's recollections." The lived experience researchers had an essential role in establishing rapport with interviewees: "Through their presence and lived experience, researchers added authenticity to the project and with considered questions, they reassured interviewees." Roslyn states: "Since working as part of this tribe I have revised my own practice." Eloise relates how even transcribing these interviews became an intimate experience: "Especially when the sound quality is quite good, you feel as if you're right there, even though [the interview] happened some time ago. It's almost like being suspended in time. Sometimes I felt like a voyeur looking in, as if I was allowed into this very intimate discussion but the interviewee didn't know that I was there."

After most of the interviews had been conducted, Roslyn checked in with several interviewees by asking them about their experience of being interviewed. Pioneering consumer activist Janet Meagher told her that being interviewed was "incredibly empowering and personally validating." As she explained:

> To have an interview that is solely about one's experiences and personal life … and having the privilege of explaining those things and having someone interested in your learnings from that part of your life, and interested in your extrapolation of experiences to inform some learning, was absolutely exhilarating in a way that meant we were valued. … A secret part of who we are has moved to a new place in the memory. (Burge, 2022, p. 155)

Several other interviewees reported being excited and feeling empowered. A few felt that the "free floating" nature of the interview was somewhat unsettling.

For the conventional academics, the nature of the team meetings and getting used to co-production provided interesting and fruitful challenges. Bellingham et al. (2022, p. 8) view co-production as a form of "slow scholarship," as designing good research projects and making decisions collectively takes time (see also Berg & Seeber, 2016). Co-production makes research much more personal as well. As Georgia reflects: "what would be lost if the research process had proceeded like a scientific protocol; the role of each team member was erased from the research in the process of writing it up." Anthony recalls that, to him, the research "at times felt unfocused, and I drifted away … [it] also delayed my engagement with the material." Others started to question common academic approaches to research that requires them to bracket their own life experiences and emotions. For Paul, the research project came at a "seminal time in my life, after a difficult divorce and a personal reckoning." His involvement resulted in two important insights. First, and "most fundamental, was the gradual discovery of the blurred lines between outsider and insider scholars," which convinced him of the validity of "fraying the boundaries between the two sides of the co-production endeavour." It also inspired Paul to adopt "more participatory methods" and including "personal stories in publications." Cathy, located in Newcastle with her postgraduate student Robyn, at times felt that she was working at a distance from the Sydney-based team members. The online meetings made it much easier to be involved with the other members of the research team during those COVID years, which created a sense of community unparalleled in her past teams.

Gemma reflects that, during the research project, she was increasingly able to integrate various parts of her personality. As she puts it: "My background as an empirical scientist and analytic philosopher has left only a little bit of space for the social, the vague, the personal, and the intangible." Gemma is a deeply spiritual person, but at the university this side of her was not apparent. She has experienced episodes of mental distress and experiences hearing voices, which is not distressing to her. Consequently, "negotiating the space between my 'academic self' and my 'lived experience self' was often a tightrope." During the research project, she "was able to formally integrate it [her spiritual and lived experience selves] into my research, something which still leaves me with a vague feeling of discomfort or worry about academic rejection." Apart from developing and embracing new ways of conducting research, "working with the group challenged my expectations of personal perfection and 'normalcy' in ways I'm only beginning to appreciate." This had significant ramifications for her personally but also for her ideas on how to conduct research and what it means to be an academic.

The way we conducted our research was novel and challenging in the context of historical research. The historians in our group comment that this work did not match our earlier experiences conducting research, even appearing at times to violate scientific principles, which dictate that our personal self, interests, inclinations, and emotional reactions should be bracketed as much as possible during the research process so that they cannot cloud their objectivity. Gemma was "struck by [the group's] discussion of what is 'real' and our combination of reality and creativity. I think this is, ultimately, what we were aiming for in our research space. A place where we could be real, be human. And a place where we could be creative about what it means to be an academic and do academic work in meaningful ways." Gemma, then, is somewhat equivocal in answering whether we were successful in living up to those ideals: "there is more work to be done before this is a question I can answer." Yet, at least, the project pointed to new ways of conducting research and new ways of being a researcher.

For Samantha, the research project challenged her preconceptions of academic research and came to be a deeply personal experience. As a transgender woman, she has experienced epistemic injustice: "Some of my formative experiences were of having my spontaneous behaviour labelled as deviant, and my subjective experiences labelled delusional." During her undergraduate education in psychology and STEM, she attempted to "distance herself from my embodied identity to achieve a position of neutral objectivity," something implicitly expected of aspiring scientists. Samantha relates how she was "closeting my mind and fractured myself between two worlds. I forced myself to disavow what I knew of the world for the sake of what others could know." She continues by stating: "The transposition of disinterest for personal involvement has not only unfettered my work but also reshaped my preconceptions about what an academic is and how an academic should work." Her contact with Ruah, herself a trans woman, helped her to gain the confidence to combine parts of herself. From Ruah, she learned "that my work and insight could be of value not despite my difference but because of it. This lesson was a precious gift."

For one member of the research team, our team meetings gained additional significance after she started to have convulsive seizures for the first time. She had no prior experience or family history of epilepsy; treating physicians tried a battery of unsuccessful investigations to locate the cause. The sudden loss of control was deeply unnerving: "even with medication, I lost my independence in many ways: I was prohibited from driving, from club sports, and even from having a bath unsupervised!" She commented that she found herself "empathising with many of the critiques the lived experience researchers levelled against the services they had received. I gained a new understanding of the frustrations, powerlessness, and enduring failures of the health system." As the mental fog lifted, this team member returned to the research group and related: "All I could do was to sit and listen. I was barely tolerating the strong side effects of the high doses of medication; a trade-off I hoped would finally control the seizures and reduce the burden of worry and care placed on my family and partner. I still felt that I was valued; the group made sure that I could still engage and contribute. I think that was the best thing they did for me. It was not articulated explicitly. Ever. It was just implied. It was a form of support that I didn't think I needed, but it turned out to be the strongest support I received."

Practical matters presented a challenge: as noted earlier, our research was shaped within an academic institutional structure and governed by rules about research expenditure. The members of the research team experienced one major disappointment: the funding unexpectedly ran out several months before we had expected it. We were therefore left on a "cliffhanger," as Georgia put it. Asha reflects: "You're quite involved in co-production, but then the ending had no co-production. … co-production can't just be the start and the middle. It has to be the end as well." For them, the sudden ending was the moment the project "fell down," resulting in experiencing "a very big power differential, especially for us lived experience researchers." Not only was the unexpected ending deeply disappointing, but it made differences in power and status very explicit. The conventional researchers were not affected in any material way, as they held continuing positions. As Eloise noted: "The academics continue on, they've still got access to all the work that's been done, which they can use in another grant and make of use again if they want, but the consumers were without a job. And that was the same for me … this part of my income was no more." For her, this highlighted the "privilege and security of one group versus the potential vulnerability and financial insecurity of the other." According to her, co-production research has the potential of being exploitative: "the academics have ongoing benefits … and the people with the lived experience don't." During the project, it appeared that we had been able to manage power differences if these were put to the side; the team worked well together as long as these base realities were bracketed.

Conclusions

Our research project has navigated questions of authorship and inclusion to offer a new contribution to the history of mental illness and community psychiatry. In doing this work we have abandoned the principle that single historian-scholars are able to investigate all relevant perspectives adequately to incorporate them in sole-authored articles and monographs. Instead, we advocate inviting as many relevant interested parties as possible to engage with our research projects and inform the outcomes. Conducting research is always a collective endeavour; acknowledging everybody's involvement reveals this collective nature. Creating partnerships to conduct research therefore requires leaving traditional notions of authorship behind. All research involves countless collaboration, multiple contributions, the sharing of contacts and experiences, and the provision of forms of training and assistance, designating a single individual as author is problematic. It can mean strategically assigning exclusive credit for research outputs. Claiming sole authorship can be seen as a form of appropriation buttressed by differences in institutional power.

Recently, Scholz et al. (2023) invited everybody involved in co-production research to move beyond mere consumer involvement and embrace lived experience leadership. Yet Rose and Kalathil (2019) have criticised co-production as untenable because power differences between consumers and professionals remain while research institutions continue to privilege certain types of knowledge of others. This is witnessed in the difficulty of crafting applications for external research funding or for human ethics committees to review and assess risk. More recently, human ethics committees have been starting to address the role of lived experience in the assessment and review of projects. This is reflected in the revised Australian National Statement on Ethical Conduct in Human Research (National Health and Medical Research Council et al., 2023), which refers to "justice" concepts (p. 10, s. 1.4). In our case, after including emerging researchers and peer workers evolved as the project commenced, we found creative ways to address their payment and involvement, constructed the research as a collective endeavour, discussed the nature of research outputs, and worked to break down hierarchies of research leadership in the team, which at times was a humbling experience. As demonstrated in this chapter, the experiences of the team members inspired rich reflections on the nature of co-production research in practice. As far as the members of our team are concerned, after being part of the described research project, we do not want to return to more conventional ways of conducting research on the recent history of mental health services. Nevertheless, co-production remains an aspiration, however imperfect we were in practising it. We consider the demand "Nothing About Us Without Us" valid, and intend, in our future projects, to invite lived experience researchers, however difficult, paradoxical, or "mad" this might be.

Acknowledgements Our research project was funded by ARC Discovery grant DP19013655: The Development of Australian Community Psychiatry. We obtained approval from the Human Research Ethics Administration at the University of Sydney (protocol 2020/029). For more information on the research project and its outputs consult the research website *Re;Minding Histories* at https://www.remindinghistories.net/.

ARC Project on the History of Australian Community Mental Health

Group Agreement

When working together we agree to:

Share the air: move back if you talk a lot or have more power, move up if you don't talk a lot or have less power.

Make space/be respectful of multiple perspectives.

Pay attention to time—equal division.

Try our best to stay on topic.

Understand where people's passions are/share where your passions are.

See difference as opportunity rather than threat.

Take time to learn from each other.

Learn from each other as equals.

Interact with mutual respect.

Stay open and curious (get off the soapbox).
Practise honesty—it is ok to disagree.
Have difficult conversations when required.
Practise clear communication.
Support each other to communicate in ways that feel safe for everyone.
Be curious.
Make space for big feelings when they arise.

ARC Project on the History of Community Mental Health in Australia

Interview Guide

I. **Before the Interview**

1. Determine who is going to be interviewed (consult the list of people).
2. Find out as much as you can about the person you are going to interview (publications, grey literature, activities, etc.) so that you are well-informed during the interview.
3. Establish contact (following the protocols of the Ethics Approval): in person, by email, by calling. Preferred method is that a person known to the interviewee establishes contact first.

   If someone has nominated themselves, this process will be easier.

   [Ethics guidelines: only email or call the person once. Lack of response is probably an indication that the person does not want to be interviewed.]
4. First longer telephone call

   a. Explain the purpose of the study in detail, what is expected of the interview, how long it will last. Mention that a support person is welcome. Mention they can bring any material from the period that they might still have (photo, newsletter, notes, etc.).
   b. Discuss the time frame: most interviews are expected to take 1–2 hours. Two separate meetings are also possible of shorter duration.

c. Discuss recording the interview and ask how the interviewee feels about that. At this point, the interviewee does not have to indicate yes/no.
d. The interviewee can nominate a location, but ideally it should not be noisy. The interview can take place by Zoom as well.
e. Discuss any specific preferences re: interviewers: numbers, gender, role in the research project.
f. For consumers only: discuss payment. Rate: $**/hour.

5. Send the *Participant Information Statement* to the interviewee and the *Informed Consent Form* by email. The interviewee can read them over in their own time; this does not need to happen during the interview.

II. **The Interview**

1. **Preparation**

With name, time, and location noted, make all practical preparations:

a. Take everything you need:

- Voice recorder. Check the battery and take additional batteries. Take an empty memory card if necessary. Earplugs or headphones to gauge the audio input; if your recording device needs electricity … take the necessary cables and an extension cord.
- Set recording levels at WAV 24/48KhZ for optimal input.
- Take two copies of the *Informed Consent Form* and one copy of the *Participant Information Statement* with you. Take information about the research project with you as well. Put your name and contact details on the PIS.
- Notepad and pen.
- Money for coffee, cookies, etc. (keep receipts)

2. **Arrival (5 minutes early)**

a. Establish rapport, make sure the interviewee is comfortable.

Ask: How are you doing?

Ask: How much time do you have available for the interview? Would you like a break time halfway?

b. Reiterate the general purpose of the research project and the goal of the interview you are about to conduct. Make sure that they are understood well.
c. Ask if there are any questions.
d. Discuss:

- the interviewee can stop the interview at any time, either temporarily or altogether.
- at any time, the interviewee can request that all data associated with them be deleted.
- The interviewee can state "off the record" when providing information. Either stop the recording for as long as necessary or inform the interviewee that although these comments will be retained on the recording, they will not appear in the transcript and will not be used in any way in our research project.
- Mention that all interviews will be transcribed, and that the interviewee will be invited to review the transcript and make any changes they like if they would like to do so.
- Inform interviewee that you need to check recording levels to increase/decrease the volume IN or OUT to ensure optimal recording. Ensure the recording device is in best spot to capture both voices.
- Turn the recorder on!
- Conduct a short practice with simple question: … *what did you have for breakfast … tell them what you had for breakfast …* play it back (with headphones—unnecessary to play to interviewee) to be sure the recorder has sound input at optimal levels. Stop recording after test. It is not necessary to include the test in the recording (it can also be deleted later).

3. **Go over the Informed Consent Form (5 minutes)**

a. Can the person's name be used in research outputs? Yes/no.
b. If not: ask if the interviewee has a preferred pseudonym/?

c. How should their role or profession be indicated (i.e. volunteer, psychologist, mental health advocate, etc.)? How should the area of their current or former activity be indicated (i.e. Melbourne, the State of Victoria, all over Australia)?
d. We are looking to place the transcripts of these recordings in an archive that is publicly accessible. Does the interviewee consent to this? If these plans are realised, we will contact the interviewee again.
e. Would the interviewee like to receive further information about the study.
f. The interviewee needs to sign the Informed Consent Form before the interview can start. Store the signed copy of the Informed Consent Form; the interviewee should keep a copy as well, with their answers to the various questions marked.

4. **Voice Recorder**
   Introduce the voice recorder (or the recording function on Zoom). Ask the informed consent questions again when the recording has started. State again that the recorder can be turned off at any time.
5. **Conduct the Interview (1 hour 40 minutes)**
   Record introduction to the interview: *This is an interview with Bill Smith, who is being interviewed for the History of Australian Community Mental Health Project.*

   The interview is taking place at his home in Blaney (or in a coffee shop in Cronulla) on DATE. The interviewer is Mary Jones.

   The aims of the study are as follows and should guide the process of inquiry throughout the interview. We are interested in the following:

- How did community mental health get started/what set the story in motion? (movements, movement leaders).
- What happened? What initiatives were undertaken? What was the experience like to be involved? Who were key people (for mental health service leaders/advocates and people with lived experience of mental distress)? I wonder what it was like to be there? From your perspective, what was happening in the world/your world at the time?

- What were the outcomes? (For who did it work? For who did it not work? What innovative ways of working did you observe? Are they still present? If so, where? If not, why?)

Important during the interview are the following:

a. Engagement: Getting to know the interviewee in the present by asking them about what they do during the day, work, hobbies, etc.
b. Ask them some general background information if appropriate (vibes).
c. Ask them to discuss any materials they brought and the meaning that is associated with them.
d. Ask the person "I'd like you to tell me the story of your involvement in activism and community mental health? Can you start from the beginning, perhaps when you first encountered the mental health system?"
e. Carefully guide them through the relevant history, bringing them back if they deviate from the aims of the study.
f. Stop and get more details on specific plots, characters, or discussions that relate directly to the research aim focussing on the aims above.
g. Use non-directive prompts and ask questions on issues you are interested in related to the research.
h. It can help to plot a written timeline simultaneously with dates and events so they can anchor the narrative later.
i. Avoid your own monologues, comments, statements, etc. Your prime aim is to listen, bear witness, prompt, guide, but not engage in two-way conversation.
j. Keep time/ask them to move forward if necessary. Be gentle but assertive.
k. If the person becomes distressed: check in with them, ask them how they are, offer a short break if appropriate.
l. Ask the interviewee for suggestions about the research project: what topics should we cover? what should we stay away from?
m. Ask for suggestions of other individuals to be interviewed.

6. **After the Interview (10 minutes)**

   a. Ask the interviewee how they felt about the interview; what they think the most important points were.
   b. Ask them if they would like a copy of the audio file to be sent to them.
   c. If the person expresses distress or seems in distress offer support, both in that moment and, if appropriate, remind of support services available through

      - Lifeline on 13 11 14
      - Beyond Blue on 1300 22 46 36
      - MensLine Australia on 1300 789 978
      - Suicide Call Back Service on 1300 659 467

   d. Express your deep and sincere gratitude that your interviewee made time available. State we will in contact in 2–4 weeks regarding the transcript and to stay in touch.

Examples

- Can you tell us a bit about your background and education, and your work history?
- How did you become involved in Organisation XXX? How did you become interested in YYY.
- How did organisation XXX get established? Who was involved?
- Tell us about your role in the organisation.
- What kind of initiatives did your organisation undertake, and how did that change over the years?
- Can you tell us a bit about your work/activism/involvement?
- What were your opinions back then, and today, about the involvement of consumers?
- What other initiatives in community mental health did you observe? What did and do you think about them?

III. **After the Interview**

1. Debrief, with fellow interviewer or someone else. Discuss
   - What do you think were the most important points that were discussed?
   - What difficulties did we encounter during the interview?
   - What can be done better during the next interview?
2. Write a short summary for all to share. Place it in the Dropbox. Upload the recording of the interview.
3. Ignore all auto-prompts by recording equipment to convert the recording to MP3 … upload in WAV 24/48. It can be converted later to MP3 to send to interviewee if requested.
4. Did the interviewee suggest names of individuals who should be interviewed? Please add them to the list with a short reason why.

References

Adema, J. (2021). *Living books*. MIT Press.

Atlas, D. (2020, August 6). Co-authorship as a storytelling tool. https://www.torch.ox.ac.uk/article/co-authorship-as-a-storytelling-tool

Bellingham, B., Elder, E., Foxlewin, B., Gale, N., Rose, G., Sam, K., Thorburn, K., & River, J. (2023). *Co-design Kickstarter*. Community Mental Health Drug and Alcohol Research Network. https://mhcc.org.au/wp-content/uploads/2023/06/CMHDARN-Co-design-kickstarter-FINAL.pdf

Bellingham, B., Foxlewin, B., Rose, G., & River, J. (2022). *Co-production Kickstarter*. Community Mental Health Drug and Alcohol Research Network. https://cmhdaresearchnetwork.com.au/resource/co-production-kickstarter/

Beresford, P., & Russo, J. (2022). *The Routledge international handbook of mad studies*. Routledge.

Berg, M., & Seeber, B. K. (2016). *The slow professor: Challenging the culture of speed in the academy*. University of Toronto Press.

Biagioli, M., & Galison, P. (2013). *Scientific authorship: Credit and intellectual property in science*. Routledge.

Brett, J., Staniszewska, S., Mockford, C., Herron-Marx, S., Hughes, J., Tysall, C., & Suleman, R. (2014). Mapping the impact of patient and public involvement on health and social care research: A systematic review. *Health Expectations, 17*(5), 637–650. https://doi.org/10.1111/j.1369-7625.2012.00795.x

Bronwyn, L. (2010). Material histories in Australia and New Zealand: Interweaving distinct material and social domains. *History Compass, 8*, 805–816. https://doi.org/10.1111/j.1478-0542.2010.00716.x

Burge, R. (2022). Reflections about the impact of interviews for interviewees and interviewers in the ARC Project: The Development of Australian Community Psychiatry after 1970. *Health and History, 24*(2), 152–160. https://doi.org/10.1353/hah.2022.0022

Clark, A., Berger, S., Hughes-Warrington, M., & Macintyre, S. (2018). What is history? Historiography roundtable. *Rethinking History, 22*(4), 500–524. https://doi.org/10.1080/13642529.2018.1528046

Coleborne, C. (2006). "His brain was wrong, his mind astray": Families and the language of insanity in New South Wales, Queensland, and New Zealand, 1880s–1910. *Journal of Family History, 31*(1), 45–65. https://doi.org/10.1177/0363199005283009

Coleborne, C. (2015). *Insanity, identity and empire: Immigrants and institutional confinement in Australia and New Zealand, 1873–1910*. Manchester University Press.

Coleborne, C. (2018). Madness uncontained. In J. M. Kilty & E. Dej (Eds.), *Containing madness: Gender and 'psy' in institutional contexts* (pp. v–viii). Palgrave Macmillan.

Coleborne, C. (2020). *Why talk about madness? Bringing history into the conversation*. Palgrave Macmillan.

Coleborne, C. (2021). Coda: Speaking madness: Word, image, action. In R. Ellis, S. Kendal, & S. J. Taylor (Eds.), *Voices of madness* (pp. 403–412). Palgrave Macmillan.

Coleborne, C., & MacKinnon, D. (2011). *Exhibiting madness in museums: Remembering psychiatry through collection and display*. Routledge.

Coleborne, C., & The Waikato Mental Health Research Group (Eds.). (2012). *Changing times, changing places: From Tokanui to mental health services in the Waikato, 1910–2012*. HalfCourt Press.

Dunlop, R., & Pols, H. (2022). Deinstitutionalisation and mental health activism in Australia: Emerging voices of individuals with lived experience of

severe mental distress, 1975–1985. *History Australia, 19*(1), 92–114. https://doi.org/10.1080/14490854.2022.2028559
Edington, C. (2019). *Beyond the asylum: Mental illness in French colonial Vietnam*. Cornell University Press.
Ellis, R., & Coleborne, C. (2022). Co-producing madness: International perspectives on the public histories of mental illness. *History Australia, 19*(1), 133–150. https://doi.org/10.1080/14490854.2022.2028558
Elroy, A. (2017). *The impact of co-production: From community engagement to social justice*. Bristol University Press.
Foucault, M. (1998 [1969]). What is an author? In D. Preziosi (Ed.), *The art of art history: A critical anthology* (pp. 299–314). Oxford University Press.
Foucault, M. (2006 [1961]). *History of madness* (J. Murphy & J. Khalfa, Trans.). Routledge.
Fricker, M. (2007). *Epistemic injustice: Power and the ethics of knowing*. Oxford University Press.
Ginzburg, C. (1980). *The cheese and the worms: The cosmos of a sixteenth-century miller*. Routledge & Kegan Paul.
Goffman, E. (1961). *Asylums: Essays on the social situation of mental patients and other inmates*. Anchor Books/Doubleday.
Gorman, R., & LeFrançois, B. A. (2018). Mad studies. In B. M. Z. Cohen (Ed.), *Routledge international handbook of critical mental health* (pp. 107–114). Taylor & Francis.
Green, A., & Troup, K. (Eds.). (1999). *The houses of history: A critical reader in twentieth-century history and theory*. Manchester University Press.
Jasanoff, S. (Ed.). (2004). *States of knowledge: The co-production of science and social order*. Routledge.
Jeffs, S. (2022). Reflections of a public loony. *Health and History, 24*(2), 174–190. https://doi.org/10.1353/hah.2022.0028
Jeffs, S., & Leggatt, M. (2020). *Out of the madhouse: From asylums to caring community?* Arcadia.
Joseph, A. J. (2019). Constituting "lived experience" discourses in mental health: The ethics of racialized identification/representation and the erasure of intergeneration colonial violence. *Journal of Ethics in Mental Health, 10*(1), 1–23.
LeBlanc, S., & Kinsella, E. A. (2016). Toward epistemic justice: A critically reflexive examination of 'sanism' and implications for knowledge generation. *Studies in Social Justice, 10*(1), 59–78. https://doi.org/10.26522/ssj.v10i1.1324

McCarthy, A., Coleborne, C., O'Connor, M., & Knewstubb, E. (2017). Lives in the asylum record, 1864 to 1910: Utilising large data collection for histories of psychiatry and mental health. *Medical History, 61*(3), 358–379. https://doi.org/10.1017/mdh.2017.33

Menzies, R., LeFrançois, B. A., & Reaume, G. (2013). *Mad matters: A critical reader in Canadian mad studies*. Canadian Scholars.

Munslow, A. (Ed.). (2012). *Authoring the past: Writing and rethinking history*. Routledge.

National Health and Medical Research Council, Australian Research Council, & Universities Australia. (2023). *National statement on ethical conduct in human research*. National Health and Medical Research Council.

Nicholls, E. J. (2022). The witness seminar: A research note. *Qualitative Research, 22*(1), 166–173. https://doi.org/10.1177/1468794120974153

Perks, R., & Thomson, A. (2016 [1998]). *The oral history reader* (3rd ed.). Routledge.

Pols, H. (2001). Divergences in American psychiatry during the Depression: Somatic psychiatry, community mental hygiene, and social reconstruction. *Journal of the History of the Behavioral Sciences, 37*(4), 369–388. https://doi.org/10.1002/jhbs.1066

Pols, H. (2007). Psychological knowledge in a colonial context: Theories on the nature of the "native" mind in the former Dutch East Indies. *History of Psychology, 10*(2), 111–131. https://doi.org/10.1037/1093-4510.10.2.111

Pols, H. (2010). 'Beyond the clinical frontiers': The mental hygiene movement in the United States. In V. Roelcke, P. J. Weindling, & L. Westwood (Eds.), *International relations in psychiatry: Britain, Germany, and the United States to World War II* (pp. 111–133). Rochester University Press.

Pols, H. (2020, July 24). The future of mental health care in Indonesia. *Inside Indonesia* (141). https://www.insideindonesia.org/the-future-of-mental-health-care-in-indonesia

Pols, H., & Grace, R. P. (2022). Cartoons by Australian consumers of mental health services: Fighting stigma, questioning normality, and presenting alternate perspectives. *Health and History, 24*(2), 32–57. https://doi.org/10.1353/hah.2022.0041

Pols, H., Setiawan, G. P., Marchira, C., Irmansyah, E. S. T., Good, M.-J. D., & Good, B. J. (Eds.). (2019). *Jiwa sehat, negara kuat: Masa depan layanan kesehatan jiwa di Indonesia [The future of mental health services in Indonesia]* (2 vols). KOMPAS.

Portelli, A. (1981). The peculiarities of oral history. *History Workshop, 12*, 96–107. https://doi.org/10.1093/hwj/12.1.96
Portelli, A. (1991). *The death of Luigi Trastulli and other stories: Form and meaning in oral history*. State University of New York Press.
Porter, R. (1985). The patient's view: Doing medical history from below. *Theory and Society, 14*, 175–197. https://doi.org/10.1007/bf00157532
Porter, R. (1987). *Mind-forg'd manacles: A history of madness in England from the Restoration to the Regency*. Athlone.
Re;Minding Histories Research Group. (2024). *Re;minding histories*. https://www.remindinghistories.net
Roper, C., Grey, F., & Cadogan, E. (2018). *Co-production: Putting principles into practice in mental health contexts*. University of Melbourne. https://healthsciences.unimelb.edu.au/__data/assets/pdf_file/0007/3392215/Coproduction_putting-principles-into-practice.pdf
Rose, D., & Kalathil, J. (2019). Power, privilege and knowledge: The untenable promise of co-production in mental "health". *Frontiers in Sociology, 4*(57). https://doi.org/10.3389/fsoc.2019.00057
Rothman, D. J. (1971). *The discovery of the asylum: Social order and disorder in the new republic*. Little Brown.
Scholz, B., Stewart, S., Pamoso, A., Gordon, S., Happell, B., & Utomo, B. (2023). The importance of going beyond consumer or patient involvement to lived experience leadership. *International Journal of Mental Health Nursing, 33*, 1–4. https://doi.org/10.1111/inm.13282
Scull, A. (1979). *Museums of madness: The social organization of insanity in nineteenth-century England*. Allen Lane.
Smart, G. L., & Zappa, A. (2022). (Re)writing care: Critical histories of community mental health services in Australia and New Zealand. *Health and History, 24*(2), 1–4. https://doi.org/10.1353/hah.2022.0039
Tansey, E. M. (2007). Witnessing the witnesses: Potentials and pitfalls of the witness seminar in the history of twentieth-century medicine. In R. E. Doel & T. Söderqvist (Eds.), *The historiography of contemporary science, technology, and medicine: Writing recent science* (pp. 276–294). Routledge.
Tomes, N. (1994). *The art of asylum keeping: Thomas Story Kirkbride and the origins of American psychiatry*. University of Pennsylvania Press.
Watts, A. (2023). Maternal insanity in the family: Memories, family secrets, and the mental health archive. *Genealogy, 7*(1), 5. https://doi.org/10.3390/genealogy7010005

6

Caring Madly/Madly Caring

Aimee Sinclair

Introduction

To speak of care within the context of mental health is risky.

I wrote that line sitting outside in the sun, at a coffee shop on the banks of Derbarl Yerrigan (colonially known as the Swan River in Western Australia), where I meet regularly to write with a colleague. Sitting there I could momentarily intellectualise care, thinking about it from a disembodied perspective, from a privileged position, as if care, or lack of care, didn't really affect me, or those around me. But as I drove home, I felt discomfort rising. Something was twisting my insides. Then the tears flooded me. To speak of care within the context of mental health is not just risky. It's heart-wrenching.

A. Sinclair (✉)
Curtin University, Bentley, WA, Australia
e-mail: aimee.sinclair@curtin.edu.au

N. A. Hendry, E. Karageorgos (eds.), *Critical Mental Health in Australia and Aotearoa New Zealand*, The Politics of Mental Health and Illness,
https://doi.org/10.1007/978-3-031-82861-4_6

For individuals such as myself, journeying through or alongside experiences commonly labelled as mental illness, care is often synonymous with coercion, seclusion, restraint, neglect and epistemic injustice (Daya, 2022). Amidst drafting this chapter, Edwige Nsilu died whilst being 'cared for' on a UK mental health ward, left unconscious for ten minutes because staff assumed she was "faking unconsciousness" (Cuddeford & Shaw, 2023). The hospital's response to her death was to no longer provide 'care' for individuals with certain diagnoses. As I sat editing, I flicked open my browser to see an Australian article speaking to incidences of sexual assaults for women, often heavily sedated, whilst in mental healthcare (Miller, 2023). Such stories are not unique; there are dozens of reports and articles speaking of unbearable harm and neglect in the name of mental health care. So much so, that many individuals, including myself, call ourselves 'survivors', not survivors of distress, but rather survivors of psychiatric care.

As the COVID pandemic was breaking, I sat, sleep-deprived, breastfeeding my four-month-old whilst my mother, my biggest advocate, my safe place in the world, revealed she had only months to live. Between that time and her passing, I needed to be cared about, and for, whilst I also cared. There were moments where living felt impossible, and I would otherwise have been (re)diagnosed mentally ill. Yet I did not ring suicide help lines, for fear of police showing up, for fear of being taken away whilst my baby cried out for me. I did not accept a bed at a mother-baby mental health unit, for fear of my autonomy being taken away. For fear of being forced on to medications that had taken me years to withdraw from. For fear I would be solely responsible for my little one, in an unfamiliar environment, without my own relations of care, however lacking they felt at the time. For fear, having been in mental health units before, I might never find my way back. I resisted talking to my psychologist, knowing such reports could be submitted at a court hearing, used to suggest I might be an unfit mother. Those mad, of course, are considered unable to care, harmful even (Sweeney & Taylor, 2021). I rarely sought out care from friends or family, cautious about people knowing the depth of my madness. Some wanted to relieve me of my caring duties. Yet I wanted to care; it was one of the few things keeping me alive.

Within psychiatric care, my distress, or madness, as I identify it, is conceptualised as an individual and biological condition; a faulty brain or psyche requiring intervention. Care, under psy-logic, equates to treatment to cure one's flawed being, or incarceration, for those who cannot be cured (Daya, 2022). Alternative imaginings of madness are obscured; as a reasonable or valid response to environmental relations, or as a potentially valuable resource for highlighting what is un-liveable within the world. Also foreclosed is the idea that I may care for myself and others through, or because of, my madness; that one may give and receive care simultaneously. My circumstances at the time meant I had sufficient privilege to avoid state-based violence: to purchase childcare, meals, cleaning, a decent doctor and therapist. Enough privilege to know what to say and not to say. A partner open to challenging gender norms and a mother, who despite being terminally ill, continued to care. Still, the threat of involuntary treatment and/or my child being removed from my care loomed. My avoidance of psy-care: a symptom of 'illness'.

Speaking and thinking of care within the context of mental health often floods me with such memories and their associated bodily responses; affects related to the absence of care, or care as oppressive. In this chapter, I draw on such affects, not as something to be avoided, but rather as resources to think through, and beyond, psy-care. I read these affects through/against the stories, affects and practices of others, who, like me, have experience facilitating peer support within the Australian mental health system.

Peer support work, as a relatively new discipline within the Australian mental health system, involves drawing on collective expertise acquired from having lived/living through distress or madness to walk alongside others navigating such experiences (Meagher & Naughtin, 2018). Peer support challenges psy-logics that enact individuals labelled mentally ill as lacking expertise and ability to care for self and others. Rather, peer support asserts that through the "experience of being socially, politically, and materially shattered", individuals develop invaluable skills and expertise in navigating distress and systems of care (Eales & Peers, 2021, p. 166). With roots in the consumer/survivor/ex-patient (C/S/X) movement, a social justice movement led by, and centring, communities oppressed by psychiatric care, peer support recognises and validates experiences of structural oppression, marginalisation and exclusion, including the effects of absent, exclusionary and/or harmful mental health care (Bakshi, 2021).

As a result of collective harms under the guise of 'care', peer support discourses tend to reject the concept of care, opting instead for 'support' (McLaughlin, 2020; Scott & Doughty, 2012). As with broader disability politics, the right to a supported independent life has been a central tenant of C/S/X activism, challenging assumptions of individuals as dependent and burdensome, lacking capacity for self-determination. Within such a context, peer support is enacted as an alternative to psy-care, underpinned by mutuality, reciprocity and self-determination (Stratford et al., 2019). In the absence of alternative conceptualisations, to speak of care aligns with care that harms, that wrestles control from those deemed mad, and sits in opposition to peer support.

Yet I suggest we do Mad folk, and the work we do in ensuring each other's survival, a dis-service by disregarding care. Rather than disregard care as always aligning with psy-logics, in this chapter I instead propose Mad care, as relations of care with emancipatory potential, informed by recipients of psychiatric services and/or those incarcerated by services. Care that resists cure and treatment, that affirms Mad ways of thinking-being-feeling, whilst also validating, and working with, the materiality of distress and individual desires for relief.

I start my discussion with why I persist with care, despite the seemingly progressive move from 'care' to 'support'. I then outline the relations that have supported me to think about care differently, namely entanglements of my own experiences of care, discussions with peer support workers, Mad studies and post-humanist theorising. Drawing on these entanglements, in the following section I explore potentials of Mad care. Finally, I consider how the inclusion of peer support within mental health systems potentially affects the production of Mad care.

Like all psy-survivor pieces, this chapter is personal and political. In some ways, it is a love letter to my mother, who cared for me (im)perfectly, traversing landscapes of madness alongside me from a place of deep compassion and solidarity, keeping me alive, treading the delicate balance between caring for, and caring about, refusing to see me as needing cure. Writing this piece has also enabled me to channel my grief and anger regarding psy-care into valuing enactments of radical care within an otherwise hostile mental health system. It is thus also a love letter to peer support workers and fellow Mad comrades, who I suggest often care for, and about, in radical ways, not despite of, but because of, their madness. Whilst I use the language of madness as a big 'f*ck you' to the ways in which psy-practices have deemed me flawed, many of those whom I talked to for this project may not align with the language or politics of madness. I do, however, feel relatively confident in their desires for individuals they walk alongside to experience connection, belonging, safety and justice. Thus, I use madness as a lens to analyse resistances to psy-violences that constrain such experiences, rather than as an essentialist marker.

Troubling Support and Care

Care, as highlighted by many feminist scholars, produces and sustains the world, rendering life liveable. Despite aversions to dependency in modern colonial societies, feminists have argued vulnerability and interdependence are universal human conditions. Garland-Thomson (2011) argues this vulnerability, whilst universally inherent, "is a potentiality that is realized when bodies encounter a hostile environment and is latent in a sustaining environment" (p. 600). Care then, may be understood as a "species activity that includes everything we do to maintain, continue and repair our 'world' so that we can live in it as well as possible" (Tronto & Fisher, 1990, p. 40). Care within this context is vital, deeply relational, entangled and inseparable from the messiness of everyday life and our relationships with self, others and the world.

Despite these affirmations of care, whilst reviewing various seminal care texts, those uncomfortable feelings of mine stirred again. I would read a page or two, and then need to walk away. Particularly unsettling were discussions that binarised caregiver and recipient, considering care only from the 'giver' standpoint; care as an ethical disposition and benevolent practice of taking responsibility for another's wellbeing. Such framings dominate mental healthcare. Rather than maintain, continue and repair our worlds, psy-care often harms, reproducing and exacerbating psy-(and other)-violences under the guise of 'taking responsibility' for another. Thus, despite my appreciation of care as a feminist, life-sustaining practice, I find it tempting, to abandon the concept of 'care' within mental health contexts, instead adopting 'support'.

'Support', as advocated in place of 'care', has been eagerly embraced within Australian mental health policy and practice, where activist desires for self-determination have imbricated smoothly with a logic of choice. A logic of choice refers to neoliberal practices through which agency and responsibility is attributed to individuals, enacted as 'consumers', to autonomously choose supports (Mol, 2008). The emotional and relational aspects of care are bracketed out, as are practices of 'doing for', conceived as preventing individuals from 'taking responsibility', and creating dependency. In this way, logic of choice and psy-logic converge with peer support through inclusion within mental health policy and practice (Sinclair et al., 2023a, 2023b). 'Recovery', originally conceived by C/S/X activists as recovery from the harmful effects of mental healthcare, becomes cure or assimilation; "learning to change habits of thinking" for a "(white, middle class genderless) heteronormative, individualized self" (Fullagar et al., 2019, p. 202; Howell & Voronka, 2012). Carelessness becomes morally endorsed; caring prevents 'recovery'.

Within such logic, peer support becomes part of a raft of 'tools' to enable independence and autonomy (Fritsch, 2010; Mol, 2008). Peer support workers are produced as role models and service connectors, providing time-limited supports to encourage individuals "to take more control of their lives" (National Disability Insurance Scheme, n.d.) through goals aligned with neo-liberal ideals of independence and productivity (Sinclair et al., 2023a).

The representation of this move from 'care' to 'support' as universally progressive, becomes complicated when read through my own everyday/night experiences with madness. There are times when I struggle to get out of bed, to shower myself, where I cannot breathe through panic. In these moments, I desire care; not physical support, whereby someone simply steps me out of the shower and hands me a towel, devoid of any emotional connection. Nor do I desire psycho-social support, whereby someone listens to me empathetically, or tries to role model me 'skills', whilst I remain starving, unable to prepare food myself. Nor do I desire support to access 'treatment'; I know these moments will pass without intervention. In such moments I often don't feel deserving of supports and cannot reach out for them. What happens when it is my responsibility to 'choose' supports? What happens when I lack the desire to get out of bed in the morning or leave the house? Under a logic of choice, I am left to 'choose' not to get out of bed. Under a logic of choice, responsibility, "the weight of everything that goes wrong", is shifted to me (Mol, 2008, p. xi). I fail to overcome my madness, to 'recover'. There are times when I crumble under this weight, and I desire someone to care. Someone to share this weight; to work with me, to maintain aspects of my life when I can't, or when I can't figure out what I need.

Yet I do not desire the type of benevolent care that involves marking me as 'other' and 'less than'. Rather, I desire care that is emotionally, materially and politically attentive. I desire what Tronto (2020) defines as being 'cared about': the affective disposition that comes with being invested in someone, alongside being 'cared for': the often mundane and monotonous practices involved in maintaining one's life. I want care that values me, respects my knowledge, experiences and understandings of distress that exist outside of psy-logics, that refuses to try to fix me, that acknowledges the ways I care for myself, unrecognisable through psy-logic. That is, I also desire a 'caring with': an ethico-political commitment to doing care in relation.

Just as my own experiences trouble the move away from care, so too did the experiences and practices peer support workers shared with me through discussions I outline shortly. Despite relations enacting peer support as goal-orientated, impersonal transactions, our discussions were never absent of care. Peer support workers talked of caring about the people they worked with. Some spoke of love. Some spoke of small gifts exchanged, thinking of the other and what they may appreciate. Several spoke to the difficulties of peer relationships ending. The affects they described were far from the ideal 'detached' mental health professional; they experienced anger, frustration, joy, pride, grief through their interactions with peers and the mental health system. They spoke of uncaring practices and care-less environments that made it hard to enact the kind of care they wanted to. Through their work within mental health systems, distress and care—past, present and future—become folded over and into one another; embodied, entangled, haunting. Peer support workers spoke of caring in particular ways because they had an embodied knowledge of absent care, carelessness, care that is harmful. Yet they also spoke of experiencing care that made life worth living. Reducing peer practices to 'support'; a set of practical tasks or simply 'sharing one's story', misrecognises and devalues what peer support workers do. Not speaking of care would de-value what peer support workers shared with me. Failing to consider care puts peer support at risk of enacting institutional forms of care: of not caring.

As such, I choose to persist with care. I now turn to how we might imagine care otherwise.

Thinking Through Care and Madness

To think and feel my way through care differently, I have drawn on my own experiences as both a giver and receiver of care, as well as the experiences, practices and theorising of 15 individuals who, like me, have experience facilitating peer support within the Australian mental health system. As part of a research project exploring the politics of inclusion, I met online or face-face with each contributor, where we shared and theorised about our experiences of inclusion as peer support workers within the Australian mental health system.[1] Our discussions encapsulated diversity across length of peer support experience, location, educational background, diagnosis, public/private/community settings and political stances regarding mental healthcare. The focus of our conversations (which might otherwise be known as interviews within more traditional research) was not predetermined as 'care'. And yet, as I described previously, 'care' seemed ever-present in our discussions, despite often being spoken in hushed tones, acknowledging such feelings or thoughts are considered controversial within mental health contexts. Contributors were subsequently invited to a mapping workshop, where six of us collectively felt and thought our way through the material and discursive

[1] Ethics approval for this research was granted by Curtin University (approval number: HRE2019-0152), with all discussions taking place in 2021.

relations within mental health systems that produce certain kinds of 'care'. Both processes, the one-one discussions and collective mapping, have been documented elsewhere (Sinclair et al., 2023c; Sinclair & Mahboub, 2024).

Inspired by Shuster and Westbrook (2022, p. 1), who invite an attending to joy within the "lived experiences of marginalized people", I re-read and felt my way through the transcripts of these individual and collective discussions, looking for where care relations felt joyful, care-full, co-created. I searched for where they felt enlivening; enabling an imagining otherwise to the 'recovered' subject who overcomes madness as an individual body-mind condition. I read the transcripts through/against theory, my own complex feelings and thoughts around care, as well as my embodied responses to the transcripts, drawing on such responses as a resource to complicate the analysis (Bozalek & Zembylas, 2017; Fox & Alldred, 2021; Fullagar et al., 2019). I engaged predominantly with Mad studies, particularly works considering the radical potential of madness (Bruce, 2021; Eales & Peers, 2021; Piepzna-Samarasinha, 2018), as well as post humanist considerations of care (Fullagar et al., 2019; Gibson et al., 2021; Puig de la Bellacasa, 2017). For clarity, I have italicised quotes directly from discussions with peer support workers, whilst quotes from the literature are in quotation marks but not italicised.

Engaging with Mad studies enables a critical attending to power within psy-practices, as well as a reading of madness that differs from that produced through such relations (Beresford & Russo, 2021; LeFrançois et al., 2013). In conceptualising madness, I draw on Bruce's (2021, p. 6) description of madness as both a dynamic social construction and a lived reality. That is, acknowledging how we come to be labelled mad because of an "unruliness of will that resists and unsettles reigning regimes of the normal", and yet also recognising madness is also embodied; feelings, perceptions, responses that may be painful or distressful, exhilarating or euphoric (Bruce, 2021, p. 6). Valuing madness for what it may teach us about care within relations of power means resisting *mandatory* notions of overcoming or curing 'mental illness', whilst at the same time appreciating that individuals may desire, for example, relief from madness, or someone to walk alongside whilst traversing intense periods of madness.

In thinking about Mad care, I have also found useful feminist new materialism and critical post-humanist considerations of care as emerging *between* things (Puig de la Bellacasa, 2017). That is, care as produced through shifting relations of material and discursive forces; bodies, social meanings, emotions, practices, objects, places, temporalities and technologies (Fullagar et al., 2019; Gibson et al., 2021). Care emerges *through* individuals coming together, their histories, practices, emotions, desires, as well as other socio-material elements; hospital waiting rooms, risk assessment forms, name badges, biomedical discourses, mental health policy. Thinking this way has the effect of drawing one's attention to habits, the fleshiness of bodies, objects; bringing as much attention to material relations of care as is given to language components of care.

Conceptualising care as emerging between things also unsettles the subject positionings of agentic caregiver/passive receiver, imbricating more smoothly with practices of mutuality, reciprocity and collective care valued in peer support. As such, whilst I use quotes from the transcripts within my discussion of Mad care in the following section, I have chosen not to assign these to individual peer support workers. I do so as an attempt to decentre an 'agentic' worker and recognise that voice, experiences and practices of care are "shared work" (Latimer, 2018, p. 387), collectively emerging through human and non-human relations (Mazzei & Jackson, 2012).

Relations of care can affect in ways that reduce a body's capacity, closing down possibilities of difference by coding and dominating things relative to the norm (Barlott et al., 2020). Psy-relations, I have argued, do so, fixing individuals into the static entity of passive 'care receiver', whose future is limited by a curative imaginary (Kafer, 2013). In contrast, I conceptualise Mad care as produced through relations that increase a body's ability to affect and be affected. Such relations open up ways of thinking-feeling-becoming beyond psy-logic. Such entangled relations, however, are always in flux, always shifting and becoming otherwise as they intra-act with new or different forces. Thinking of care as produced through moving entanglements (desires, emotions, hauntings of past care) means relational affects cannot be known entirely in advance. What may be experienced as emancipatory by one person may not for another, or in another moment, in another context. As such, there is no one, right, unitary way to care. Rather, what is important, is a commitment to a continual noticing, questioning, experimenting and negotiation of care practices, spaces, relations within each moment of care. Thus, my desire is that my discussion of Mad care outlined below is experienced as a provocation; a springboard for considering practices and relations that hold *potential* for constituting Mad care, rather than a representation of what Mad care 'is' or should be.

Conceptualising Mad Care

What might care look-feel-sound like for those of us experiencing deep, bone-aching, skin-crawling distress or madness, who don't necessarily desire cure, but sometimes desire relief? Care that recognises the material, affective and embodied nature of madness, without mandating cure or seizing control? Care that enables and affirms a multitude of ways of feeling-thinking-becoming in this world, that recognises the ways Mad folk desire care and provide care, for both self and others, that goes otherwise unrecognised? Drawing on entanglements of peer support worker experiences, affects, Mad and critical post-humanist theorising, I turn now to considering potential elements of Mad care.

Mad Care as Imagining Otherwise

Practices of imagining otherwise involve revealing possibilities for life beyond a curative imaginary, in which the only future possible involves intervention in the form of cure or overcoming madness. In imagining otherwise, madness is not the problem, nor is it the role of a peer support worker to resolve madness. Rather, imagining otherwise involves an attentiveness to one's "*own explanations for what is going on*", "*exploring ideas and approaches*" and valuing people finding "*their own ways forward*", "*building a community and space where people can … explore their own paths as a whole*".

The presence of someone who has journeyed through, or alongside, experiences of madness, holds potential for imagining otherwise. Such 'role modelling' resists the role modelling of recovery enacted through policy and research, where peer support workers, produced as having overcome madness, are expected to role model self-management and engagement with psy-services (Sinclair et al., 2023a, 2023b). Rather, practices of imagining otherwise involve role modelling possibilities that exist alongside madness:

> *Model[ing] that you can have the sort of mental health experiences … like feeling so anxious that you could just about jump out of your skin … and to a reasonable extent incorporate them into your life … just modelling that you can … not carry on regardless, but that you can lead a reasonably fulfilling life.*

In imagining otherwise, madness is not romanticised. There is a recognition of the difficulty of living in this world with madness. There are times, as we journey alongside madness, that one may not be able to "*carry on regardless*". Yet at the same time, practices of imagining otherwise don't produce a certain kind of hope that involves overcoming madness, of returning to a 'sane' way of being. In imagining otherwise, the work required to overcome madness as demanded by psy-logic is resisted. Psy tools and practices that produce a need to overcome madness, such as goal setting worksheets and progress measurements, are "*put aside*" or "*work[ed] around*". The focus instead is on relations underpinned by the inherent value of a person.

Imagining otherwise involves both an imagining of relationship to self and madness other than deemed by psy-logics and an imagining of futures other to a curative imaginary. Doing so is a radical act amongst psy-logic that enacts madness as all-encompassing, never receding and always impeding a 'good' life. Such acts of Mad care can be vital for those of us who have been told, or treated like, our madness means we are "*less than human*" or lacking a future. Imagining otherwise involves facilitating others to "*see their own life from a different perspective*":

> *…see that they too can achieve things and have meaning and purpose in their life, that they are an important person … they are not just whatever's happening to them at that particular time.*

Mad Care as Pursuing Pleasure

Practices of pursuing pleasure involve putting aside practices of "*sitting down and talking about mental health*" and instead pursuing places and activities that produce joy:

> *[P]ut[ting] away the list of 'to do', the jobs and the appointments ... [to] do something more pleasurable ... we might drive by the beach ... just going out for coffee, ice-creams, doing something that will inspire them or going somewhere that they wouldn't normally go to.*

Pursuing pleasure together may involve abseiling, boxing, the pool, theme parks, archery. For me, a hike in the bush can sometimes restore my will to live more than any psy-care I have experienced. When we think of care as produced through non-human *and* human material relations (Puig de la Bellacasa, 2017), it allows for a recognition that a breeze in the park, the smell of basil, a waterslide can be just as life-sustaining as anything someone says or does.

Acts of pursuing pleasure put aside the idea that one must be cured or always working towards overcoming madness to become a 'productive citizen' by attending "*the jobs and the appointments*". Rather, pursuing pleasurable activities suspends this curative future, insisting that "our present-day body-minds are as important as any vision of the future" (Clare, 2017, p. 90). In doing so, Mad practices of pursuing pleasure produce all body-minds as deserving of pleasure or joy, and resists care as cure or treatment.

Mad Care as Access Intimacy

Mia Mingus (2011) describes access intimacy as

> that elusive, hard to describe feeling when someone else 'gets' your access needs ... who have an automatic understanding of access needs out of our shared similar lived experience of the many different ways ableism manifests in our lives ... Instantly, we can hold the weight, emotion, logistics, isolation, trauma, fear, anxiety and pain of access. (n.p.)

I suggest this intimacy around needs, desires, frustrations, hauntings, comes from living through experiences of 'misfitting' (Garland-Thomson, 2011), a willingness to tap into such experiences, and relations that enable one to do so. Mad care as intimacy doesn't necessarily require 'sharing' one's experiences as encouraged by the mental health system. Rather, living with madness in an unsustaining world contributes to access intimacy through an embodied knowing; otherwise described as "*intuitively understanding*" or "*a really intrinsic understanding*". There are "*almost no words for it*". Access intimacy means not having to ask to move to the quieter place in the café, or a seat with one's back to the wall, because this has already been sensed as a potential need. It's knowing the comfort of vending machine food when one is "*absolutely uncontrollably massively hungry from the meds*". Its understanding "*when you've been unwell for a very long time, making phone calls is really hard*". Mad care as intimacy is someone exclaiming to a peer support worker; "*you're the only one that knows what the f*ck I'm talking about!*"

Importantly, access needs do not always look the same, nor does one need to know another's access needs automatically to produce access intimacy. What might be a space of comfort to one, might be a place of fear to another, or to the same person the following week. Care hauntings—conflicting discourses, relational histories, experiences, reminders of others—are always entangled with care, such that the affects of mad care practices can never be known in advance. Rather, access intimacy is created through curiosity and an attentiveness to shifting desires and differences; a willingness "to be undone by another" (Eales & Peers, 2021, p. 118).

Mad Care as Doing Nothing/Doing Something

Practices of 'witnessing'; the act of "*just listening*", fostering connection and prioritising relationship over doing or fixing may produce Mad care. Such acts, sometimes described as 'doing nothing', involve "*just being there … and being calm … not talking about illness and medication and diagnosis … just having a conversation … just sitting with them … just connecting*". Doing so has the effect of giving "*the person the thought that …*

'maybe I've got something worthwhile to say ... the fact that she's listening to me makes me a somewhat valuable person". Through 'doing nothing' someone becomes of value, as opposed to, psy-care, that enacts those deemed mad as not having "*anything to offer*". Acts of 'doing nothing' can create space to "*explor[e] ideas and approaches and opportunities outside the confines of those defined illnesses*", connecting 'doing nothing' in with imagining otherwise. Doing nothing resists the idea that madness needs fixing.

Importantly, 'doing nothing' is neither neglect nor abandonment. Mad care as 'doing nothing' involves a high level of skill and expertise, "understanding and appreciation of a person's unique way of being in the world" (Spandler & Stickley, 2011, p. 559). Yet within a system that prioritises tangible effects, the skill and labour that goes into being with someone and their distress is rendered invisible, such that these acts of Mad care go unrecognised or devalued. Such practices contrast with "many patients experience of hospital wards where they feel just observed and monitored"; where 'nothing' is done (Spandler & Stickley, 2011, p. 559).

Whilst Mad care may involve 'doing nothing', it may also manifest through 'doing something'; "the more tangible things, like when people actually get out of the house, and go for a walk on the beach ... go[ing] with someone to the employment agency". 'Doing something' may involve working towards new possibilities, as well as the everyday mundane labour of maintaining aspects of life until a crisis passes. It may involve attending appointments, making phone calls, supporting someone to return bags of shopping after a spending spree.

In some ways, 'doing something' resembles caring practices often referred to as 'caring for'; the hands-on, often mundane and over-looked care labour. However, for Mad folks, someone 'caring for' has often involved assuming we can't, or refuse to, do it ourselves, assuming what is best, assuming we need 'fixing' and taking control from us. Practices of 'doing something' recognise those of us who don't necessarily want to be cured or overcome madness but may feel too broken to work through our overflowing washing basket, our piles of dirty dishes, the phone calls that seem overwhelming; tasks that another time might be done without a second thought. Mad care resists assuming we need to be taught the skills to do such things, or that we simply need motivation to do them. Mad care may be believing our expertise when we say we need someone to do these tasks for us, with us or alongside us, or when we want to do it ourselves. It refutes neoliberalist and sanist ideas that dependence leads to more dependence, asserting rather that mind-bodies are always interdependent, and all require care in various ways.

'Doing something' involves a working together to open up possibilities rather than close them down, guided by the desires, histories and temporal pacing of both individuals, and being attentive to the power relations within which we are all entangled. "*Getting to know [people] and what's important to them … what their values are … and how they want their life to be*" is a vital aspect of doing something. Rather than 'doing something' to cure or overcome madness, together we may work together to maintain and care for life. In doing so, we become other to an active caregiver/passive receiver; we produce Mad care in relation.

Such Mad acts of 'doing something' are also de-valued within the mental health system, assumed to require minimal skill and thus relegated to the lowest paid workers. Differentiating Mad care from transactional 'support', Mad care involves intentionality. That is, 'doing something' involves more than undertaking a set of practical tasks as outlined in a (often organisationally or clinically driven) support plan, but rather involves a "complex human intimate interaction which requires skill and understanding" (McLaughlin, 2020, p. 400). It involves drawing critically on collective lived experience, applying it in a skilful way, remaining attentive to shifting relations and desires. As such, Mad care is produced through a "caring for": the material practices involved with maintaining one's life or creating new futures, "caring about" from a place of solidarity, the affective disposition that comes with being invested in someone, and "caring with".

Mad Care as Creating/Becoming Together

Mad care asks "what can be created together?" (Fritsch, 2010, p. 12). Practices of creating together challenge both a logic of choice, whereby responsibility is placed on the individual to make 'good' choices, and psy-logic which involves "*taking power*" away from the individual, where "*support is done to*" someone because the worker decides "*what's best for someone else*". Mad care resists forced treatment, refusing to align with psy-practices, always aligning with those experiencing madness, always taking the side of the oppressed.

Creating together involves a refusal to provide advice, challenging the subject of peer support worker who is 'recovered' providing support for one who is not. Rather, creating together may involve sharing tips together about how one gets out of bed in the morning, without insisting that getting out of bed in the morning is necessarily the best thing for each other. It may involve brainstorming together what might be helpful, knowing that what might be helpful for one person may not for another. "*There is 'no better than you' because I'm a worker and you are an unwell person*". Rather, it's "*I've been there, I've been in your shoes*".

This mutuality within Mad care acknowledges we are all made small through systems of oppression, and that we all become anew through our interactions. In this way, creating and becoming together are interwoven. Mad care is negotiated, allowing for a sharing and affirming of both parties' desires, histories, meaning-making and practices. Such negotiations resist a logic of choice, whereby workers are expected to be a "detached tool", performing a series of mechanical tasks (Gibson, 2006, p. 192). Care becomes a shared responsibility that requires all actors to engage relationally in ways that enhance life possibilities, rather than something that requires individual effort such as goal setting and self-care strategies. Together, we may "*meet up and connect and take things slow and build a relationship and share these mutual intersecting challenges and things that have been helpful*".

Mad Care as Shared Vulnerability

Mad care as vulnerability refutes the performative nature of 'recovery' often required by peer support workers to gain employment and orientate others towards recovery, whereby recovery involves overcoming madness (Voronka, 2017). Madness, in this sense, is only valuable if it exists in the past. Challenging such notions involves conceptualising "*ongoing messiness*" as "*an asset to employ*", recognising that current experiences of madness/distress are not inherently prohibitive of care. Rather, current experiences of madness may enable "*learning from each other in the moment*", demonstrating possibilities of doing "*the things that you want whilst feeling 'unwell' sometimes*".

Sharing vulnerability with others enables a becoming together, where peers may learn from one another to create new understandings and practices:

> *If we're having a flat day, we will say we are having a flat day and our voice will be flat … they [non-peer workers] might do that in the office, but they go out and work with someone and act like everything's okay … that's the difference … we're always ourselves and we can say we're having a bad day and, you know, apologize … "I'm sorry that I'm not one hundred percent here for you today" … you can be unwell and still do things.*

Madness itself does not prevent care, but rather such capacities, such as the capacity "*to hold space for someone else*" or "*be present for someone and be able to put them first*" are enabled or constrained depending on the entanglement of madness with broader relations. One may be having a "*bad day*" or going through the "*worst times of [one's] life*" and still be able to engage in Mad care. Thinking of my own experiences, I have cared through periods of feeling suicidal, through periods where my mind-body has slowed from the effects of madness. I may not be able to talk on the phone or navigate clinical spaces, but I can care via text. I might not be able to meet institutional goals regarding how many people I need to 'support', and yet, I can sit with someone, tuned in to sensitivities and shared temporalities and desires, unable to, and thus resisting the fast-paced, goal setting practices shaping peer support.

Caring Madly Within a Care-less System

Many of the peer support workers I talked with were adamant that what I have described above as moments of Mad care occur despite, or in resistance to, the mental health system. As such, having considered potentials of Mad care, I now turn my thinking to how the inclusion of peer support within mental health systems potentially affects the production of Mad care. I contend the Australian mental health system (with its entanglements of clinical spaces, temporalities, name badges, risk assessments, organisational procedures, biomedical discourses, mental health policies) limits the ability to both imagine and practise Mad care.

For many peer support workers, it is deeply embodied experiences of harmful care, or experiences of an absence of care, that prompt a desire to work within, and resist, psy-relations of care, and inform the way they desire to work with others. Yet working within mental health systems is risky for peer support workers, with bullying, microaggressions, over-management and re(diagnosing) of mental illness common for workers who fail to adhere to psy-logic and practices, or who experience moral injury from working within such environments (Byrne et al., 2016; Edan et al., 2021). Engaging in practices of Mad care "*doesn't fit*" within the system, with "*consequences for you doing the thing*".

Further, peer support workers talked of the "*red tape*", "*caveats and restrictions*", "*hoops*" and "*regulations*" that prevented them enacting Mad practices such as pursuing pleasure. "*Sanitised*" clinical spaces, the "*generic sameness*" of a hospital room, standardised recovery and assessment tools, and organisational policies come together to produce care as transactional and "*clear-cut*" rather than relational. Similarly, despite desiring mutuality and reciprocity, entangled relations of remuneration, name badges, recovery discourses, case notes, staff-only spaces, organisational policies (and so many more), enact peer support workers as 'recovered' and thus 'better than' those accessing services:

> *Even though we're saying, 'Yeah I'm a peer worker', and there's a value of mutuality and wanting to work towards that, there is still a power imbalance.*

Within such entanglements, it becomes hard to imagine care other to psy-care. Thus, whilst the embodied affects of madness and psy-care that peer support workers bring to relationships carry potential to produce emancipatory care, there were many times during discussions with peer support workers, and when reading and feeling my way through the transcripts, where I felt smaller from descriptions of care enacted by peer support workers. I have often, in the name of peer support, enacted care with problematic effects. I am ashamed particularly, of my part in replicating colonialist practices through peer support work. Reading through the earlier interview excerpts, you may have even heard or felt psy-logic and a logic of choice entangled within and through enactments of Mad care. Peer support, through its inclusion within mental health systems, is always entangled with psy-care and a logic of choice, such that the values and practices of peer support workers "*are compromised constantly*", even when one desires otherwise. Through entanglement with psy-relations, peer support becomes at risk of replicating the same harms it works to resist (Fabris, 2013; Voronka, 2017).

Further, whilst Mad care can spark joy and a becoming otherwise, the labour of care is also work: it can be exhausting, challenging. Mad care is hard. It requires sitting with sadness, discomfort, fear and unknowingness. Without this work being valued, resourced and undertaken collectively, radical care practices are unsustainable.

The "*magic*" of peer support, or what I've conceptualised as Mad care, was described by many peer support workers as happening predominantly outside the mental health system. Some peers volunteered or founded their own groups outside of paid work where Mad care was more likely to be produced, without "*other peoples' red tape*". Several had left working in clinical/inpatient services, preferring to work in community organisations, outpatient clinics or via worker hire platforms, where they had found relations more conducive to Mad care. Some had moved away from mental health systems completely.

And so, the question, at least for me, becomes, how might peer support workers (and others), collectively work towards relations that hold potential for Mad care, without being engulfed by relations that constantly produce carelessness, or shape care into control? How might we care in ways that dismantle rather than uphold the existing system? How might we live in the in-between spaces that enable Mad care, whilst also creating spaces and knowledges outside of the system?

Throughout our discussions, we mapped entangled relations (strong desires, allies, secure employment, training ground in C/S/X or other social justice movements, peer supervision, experience, Mad theorising, affects such as "*frustration and deep sadness about the way our system still works and operates*") that enabled a noticing and questioning of dominant practices. Such relations hold potential to propel peer support workers to imagine care otherwise. Several peer support workers also talked of strategies they used to "*fly under the radar*" and "*be left alone*" to do their work, avoiding or creatively filling in case notes and clinical tools, incorporating practices of Mad care within programs where they weren't "*supposed to be doing that stuff*". Such strategies might perhaps be considered a kind of sly resistance (Mills, 2014), enabling Mad folk to exist within the system, evading consequences for themselves and other peers, whilst subverting psy-practices, and enabling diverse ways of caring. Rather than fighting against the system, such entanglements produce small ruptures, lines of flight elsewhere, that are worthy of pursual as we attempt to move away from care that harms towards life-affirming, Mad care.

Imagining Otherwise

In writing this chapter, my desire was to prompt a consideration of how, as Mad folk, we may reclaim care. I have suggested Mad care involves affirming and enabling a multitude of ways of thinking and feeling in the world that includes, rather than working to cure or overcome, madness. Mad care resists and disrupts psy-logics of care associated with surveillance, punishment and abandonment, as well as a logic of choice, where support is provisional on working towards self-sufficiency and independence. Caring madly creates possibilities beyond the pathological, unproductive or empowered 'consumer', and the detached, 'recovered', care-less support worker.

Whilst I have suggested some potential relations of Mad care, there are many more to be considered. For example, whilst Mad care centres sanism, it must be "attentive to its inherent intersections with racism, sexism, classism, heterosexism, transphobia and ableism" (Costa & Ross, 2022, p. 6). Attentiveness to these intersections is a vital part of justice-doing, yet such differences become erased under an essentialist marker of 'lived experience', the basis upon which peer support workers are employed (Voronka, 2016). This attentiveness to difference was not prevalent across our discussions yet is an important part of Mad care that should be attended to.

During my time within mental health and peer spaces, I have often heard the idea that peer support is non-political. That is, broader advocacy work is political or activist work; but peer support is not. The concept of Mad care disrupts this idea. Imagining and experimenting with alternative ways of doing care, I hope, will enable us one day to abolish carceral psychiatric institutions. When I reflect on my own experiences, the harms of psy-care still reverberate through my body and my everyday life, despite being twenty-five years since I was first labelled mentally ill and admitted into the world of psy-care. The threat of psy-care remains. It is only through privilege that I have been able to resist. I desire a world where privilege is not required to be cared for in ways that are life-sustaining. Considering how we might care madly could be one way to do so.

Acknowledgements Whilst this chapter is sole-authored, the thinking within it is collective. I am especially indebted to the peer support workers who contributed to the thinking shared here, alongside my research supervisory team: Dr Christina Fernandes, Dr Sue Gillieatt and Lyn Mahboub, and all those who have fought, and continue to fight, for the rights of folks deemed mad.

References

Bakshi, S. (2021). Peer support as a tool for community care: "Nothing about us, without us". *Columbia Social Work Review, 19*(1), 20–43. https://doi.org/10.52214/cswr.v19i1.7602

Barlott, T., Shevellar, L., Turpin, M., & Setchell, J. (2020). Destabilising social inclusion and recovery, and pursuing "lines of flight" in the mental health sector. *Sociology of Health & Illness, 42*(6), 1328–1343. https://doi.org/10.1111/1467-9566.13106

Beresford, P., & Russo, J. (2021). *The Routledge international handbook of Mad studies* (1st ed.). Routledge.

Bozalek, V., & Zembylas, M. (2017). Diffraction or reflection? Sketching the contours of two methodologies in educational research. *International Journal of Qualitative Studies in Education, 30*(2), 111–127. https://doi.org/10.1080/09518398.2016.1201166

Bruce, L. M. J. (2021). *How to go mad without losing your mind: Madness and Black radical creativity*. Duke University Press.
Byrne, L., Happell, B., & Reid-Searl, K. (2016). Lived experience practitioners and the medical model: World's colliding? *Journal of Mental Health, 25*(3), 217–223. https://doi.org/10.3109/09638237.2015.1101428
Clare, E. (2017). *Brilliant imperfection: Grappling with cure*. Duke University Press.
Costa, L., & Ross, L. (2022). Mad studies genealogy and praxis. *International Mad Studies Journal, 1*(1), e1–e8. https://doi.org/10.58544/imsj.v1i1.5239
Cuddeford, C., & Shaw, N. (2023, June 27). Woman died in hospital after nurses thought she was pretending to be unconscious. *Wales Online*. https://www.walesonline.co.uk/news/uk-news/woman-died-hospital-after-nurses-27204648
Daya, I. (2022). Russian dolls and epistemic crypts: A lived experience reflection on epistemic injustice and psychiatric confinement. *Incarceration, 3*(2), 263266632211034. https://doi.org/10.1177/26326663221103445
Eales, L., & Peers, D. (2021). Care haunts, hurts, heals: The promiscuous poetics of queer crip Mad care. *Journal of Lesbian Studies, 25*(3), 163–181. https://doi.org/10.1080/10894160.2020.1778849
Edan, V., Sellick, K., Ainsworth, S., Alvarez-Varquez, S., Johnson, B., Smale, K., Randall, R., & Roper, C. (2021). Employed but not included: The case of consumer-workers in mental health care services. *The International Journal of Human Resource Management, 32*(15), 3272–3301. https://doi.org/10.1080/09585192.2020.1863248
Fabris, E. (2013). What could go wrong when psychiatry employs us as 'Peers'? In B. A. LeFrançois, R. Menzies, & G. Reaume (Eds.), *Mad matters: A critical reader in Canadian mad studies* (pp. 130–139). Canadian Scholars' Press.
Fox, N. J., & Alldred, P. (2021). Applied research, diffractive methodology, and the research-assemblage: Challenges and opportunities. *Sociological Research Online*, 1–17. https://doi.org/10.1177/13607804211029978
Fritsch, K. (2010). Intimate assemblages: Disability, intercorporeality, and the labor of attendant care. *Critical Disability Discourse, 2*, 1–14.
Fullagar, S., O'Brien, W., & Pavlidis, A. (2019). *Feminism and a vital politics of depression and recovery*. Palgrave Macmillan.
Garland-Thomson, R. (2011). Misfits: A feminist materialist disability concept. *Hypatia, 26*(3), 591–609. https://doi.org/10.1111/j.1527-2001.2011.01206.x
Gibson, B. E. (2006). Disability, connectivity and transgressing the autonomous body. *Journal of Medical Humanities, 27*(3), 187–196. https://doi.org/10.1007/s10912-006-9017-6

Gibson, B. E., Fadyl, J. K., Terry, G., Waterworth, K., Mosleh, D., & Kayes, N. M. (2021). A posthuman decentring of person-centred care. *Health Sociology Review, 30*(3), 292–307. https://doi.org/10.1080/14461242.2021.1975555

Howell, A., & Voronka, J. (2012). Introduction: The politics of resilience and recovery in mental health care. *Studies in Social Justice, 6*(1), 1–7.

Kafer, A. (2013). *Feminist, queer, crip*. Indiana University Press.

Latimer, J. (2018). Afterword: Materialities, care, 'ordinary affects', power and politics. *Sociology of Health & Illness, 40*(2), 379–391. https://doi.org/10.1111/1467-9566.12678

LeFrançois, B. A., Menzies, R., & Reaume, G. (Eds.). (2013). *Mad matters: A critical reader in Canadian Mad studies*. Canadian Scholars' Press.

Mazzei, L. A., & Jackson, A. Y. (2012). Complicating voice in a refusal to "let participants speak for themselves". *Qualitative Inquiry, 18*(9), 745–751. https://doi.org/10.1177/1077800412453017

McLaughlin, J. (2020). Valuing care and support in an era of celebrating independence: Disabled young people's reflections on their meaning and role in their lives. *Sociology, 54*(2), 397–413. https://doi.org/10.1177/0038038519868630

Meagher, J., & Naughtin, G. (2018). Peer work in mental health in Australia: Its scope, role and contribution. In J. Meagher, F. Jackson, A. Stratford, E. Jayakody, & T. Fong (Eds.), *Peer work in Australia: A new future for mental health* (pp. 3–40). Richmond PRA and Mind Australia.

Miller, A. (2023, July 24). In Australian mental health units, women are drugged up, dehydrated and retraumatised. Now they're speaking out. *ABC News*. https://www.abc.net.au/news/2023-07-24/australian-mental-health-units-women-patients-retraumatised/102415280

Mills, C. (2014). Sly normality: Between quiescence and revolt. In B. Burstow, B. A. Lefrançois, & S. Diamond (Eds.), *Psychiatry disrupted* (pp. 208–224). McGill-Queen's University Press.

Mingus, M. (2011, May 5). Access intimacy: The missing link. *LeavingEvidence*. https://leavingevidence.wordpress.com/2011/05/05/access-intimacy-the-missing-link/

Mol, A. (2008). *The logic of care: Health and the problem of patient choice*. Routledge.

National Disability Insurance Scheme. (n.d.). Psychosocial recovery coach information. https://www.ndis.gov.au/media/2412/download

Piepzna-Samarasinha, L. L. (2018). *Care work: Dreaming disability justice*. Arsenal Pulp Press.

Puig de la Bellacasa, M. (2017). *Matters of care: Speculative ethics in more than human worlds*. University of Minnesota Press.

Scott, A., & Doughty, C. (2012). Care, empowerment and self-determination in the practice of peer support. *Disability & Society, 27*(7), 1011–1024. https://doi.org/10.1080/09687599.2012.695578

Shuster, S. M., & Westbrook, L. (2022). Reducing the joy deficit in sociology: A study of transgender joy. *Social Problems, 1–19*. https://doi.org/10.1093/socpro/spac034

Sinclair, A., Fernandes, C., Gillieatt, S., & Mahboub, L. (2023a). Peer work in Australian mental health policy: What 'problems' are we solving and to what effect(s)? *Disability & Society, 1–26*. https://doi.org/10.1080/09687599.2022.2160926

Sinclair, A., Gillieatt, S., Fernandes, C., & Mahboub, L. (2023b). Inclusion as assimilation, integration, or co-optation? A post-structural analysis of inclusion as produced through mental health research on peer support. *Qualitative Health Research, 33*(6), 543–555. https://doi.org/10.1177/10497323231163735

Sinclair, A., & Mahboub, L. (2024). Maddening post-qualitative inquiry: An exercise in collective (Mad) theorising. *Qualitative Health Research*, 10497323241231896. https://doi.org/10.1177/10497323241231896

Sinclair, A., Mahboub, L., Gillieatt, S., & Fernandes, C. (2023c). 'You just treat me like a human being': Using lived experience to (re)imagine boundary practices in mental health settings. *The British Journal of Social Work, 53*(3), 1408–1425. https://doi.org/10.1093/bjsw/bcad044

Spandler, H., & Stickley, T. (2011). No hope without compassion: The importance of compassion in recovery-focused mental health services. *Journal of Mental Health, 20*(6), 555–566. https://doi.org/10.3109/09638237.2011.583949

Stratford, A. C., Halpin, M., Phillips, K., Skerritt, F., Beales, A., Cheng, V., Hammond, M., O'Hagan, M., Loreto, C., Tiengtom, K., Kobe, B., Harrington, S., Fisher, D., & Davidson, L. (2019). The growth of peer support: An international charter. *Journal of Mental Health, 28*(6), 627–632. https://doi.org/10.1080/09638237.2017.1340593

Sweeney, A., & Taylor, B. L. (2021). De-pathologising motherhood. In I. P. Beresford & J. Russo (Eds.), *The Routledge international handbook of Mad studies* (1st ed., pp. 166–176). Routledge.

Tronto, J. C. (2020). *Moral boundaries: A political argument for an ethic of care* (1st ed.). Routledge.
Tronto, J. C., & Fisher, B. (1990). Toward a feminist theory of caring. In E. Abel & M. Nelson (Eds.), *Circles of care* (pp. 36–54). SUNY Press.
Voronka, J. (2016). The politics of 'people with lived experience': Experiential authority and the risks of strategic essentialism. *Philosophy, Psychiatry, & Psychology*, 23(3), 189–201. https://doi.org/10.1353/ppp.2016.0017
Voronka, J. (2017). Turning Mad knowledge into affective labour: The case of the peer support worker. *American Quarterly, 69*(2), 333–338. https://doi.org/10.1353/aq.2017.0029

7

Tēnei te pō nau mai te ao: Activating Indigenous Systems Change

Diana Kopua, Mark Kopua, and Michelle Levy

Introduction

There is an extensive literature base documenting the inequitable health outcomes which have resulted for Indigenous peoples as a result of settler colonialism (Axelsson et al., 2016). Māori, the Indigenous people of Aotearoa New Zealand, experience significant inequities in health compared to the non-Indigenous population (Reid et al., 2022). Whilst socio-economic deprivation is associated with poorer health outcomes, inequities remain even after adjusting for socio-economic deprivation or position (Reid et al., 2000). Unequal power relationships, the unfair distribution of the social determinants of health, biases, unexamined privilege and institutional racism are all evidenced as underpinning health inequities for Māori (Reid & Robson, 2007). Institutional racism is an accepted and known determinant of health and wellbeing in Aotearoa

D. Kopua (✉) • M. Kopua • M. Levy
Te Kurahuna Ltd, Gisborne, New Zealand
e-mail: diana.kopua@tekurahuna.com

N. A. Hendry, E. Karageorgos (eds.), *Critical Mental Health in Australia and Aotearoa New Zealand*, The Politics of Mental Health and Illness,
https://doi.org/10.1007/978-3-031-82861-4_7

(Reid et al., 2019; Waitangi Tribunal, 2019). Health systems in Aotearoa are evidenced as supporting non-Māori to live longer, healthier lives than Māori (Health Quality & Safety Commission, 2019), with Brown and Bryder (2022) demonstrating how the so-called universal health system in Aotearoa has never delivered equitable health outcomes for Māori. The severity and persistence of health inequity for Māori is itself considered an indicator that the health system in Aotearoa New Zealand is institutionally racist (Waitangi Tribunal, 2019).

The role of the health system and health professionals in creating and maintaining institutional racism and therefore inequity for Māori is becoming increasingly recognised. For over 30 years, reviews and inquiries across multiple sectors, including health, have emphasised both the failure for Māori communities of approaches premised upon Western knowledge systems and models of practice, and the need for transformative change (Boulton et al., 2020). Despite this, authentically transformative solutions which address equity for Māori remain elusive in Aotearoa. Te Whare Wānanga o Te Kurahuna (a precious source of knowledge steeped in traditional practises consistent with a Māori worldview), is the kaitiaki (guardian) of Mahi a Atua (a philosophy guided by ancestral knowledge), a uniquely Indigenous workforce development approach in Aotearoa. Deliberately aspiring to create movement to a collective critical consciousness, Mahi a Atua prioritises the reinstatement and embedding of Indigenous knowledge systems as the foundation for intentional system transformation able to address institutional racism, and therefore inequity for Māori.

First created in the 1990s, Mahi a Atua originated as a way of meaningfully engaging with Māori whānau (family group) in mental health settings. In 2014 the first two authors evolved Mahi a Atua beyond being a therapeutic practice. With the connection between institutional racism and inequity for Māori becoming clearer, Mahi a Atua began to specifically direct its attention to the creation of deliberate pathways for systemic transformation not only across health systems, but also iwi (tribal group), education, social and art sectors, and the wider community (Kopua et al., 2021).

Grounded in Indigenous ontology and epistemology, Mahi a Atua is founded on 'He Oranga Whakapapa' which acknowledges everything has an origin that can be traced back to specific pūrākau— "age-old Māori creation narratives which, for centuries, have provided a way for Māori to contextualise and find meaning in the events of everyday life and then identify culturally and spiritually acceptable pathways for their resolution" (Rangihuna et al., 2018b, p. 16). Whilst pūrākau versions vary depending on which iwi, community or individuals are sharing the story, the generic elements across the stories demonstrate both the spectrum of issues faced by our earliest ancestors, alongside the strength and resilience of the atua (deity) as they made sense of, re-balanced, and overcame struggles; in the process enacting pathways to resolution and well-being (Rangihuna et al., 2018a).

Drawing on interdisciplinary knowledge bases, including Kaupapa Māori critical theory and systems change theory, this chapter explores how Te Kurahuna operationalises an Indigenous consciousness-based approach to systems transformation. Of importance is that this approach intentionally sits within a wider Indigenous liberation movement. This movement understands the central importance of individuals and collectives being conscientised to the fact that colonisation can and is perpetrated by and against ourselves (Smith, 2017), and that through a reawakened belief in our own capacity to reflect, act and transform, communities can be supported to act transformatively via everyday individual and collective actions (Smith, 2003). Integral to this approach is a focus not only on what is required to shift conditions sustaining institutional racism, but also growing a collective leadership who are imagining and co-creating innovative Indigenous futures.

Systems Change Theory

Systems change theorists identify three fundamental conditions of change: structural change; relational change; and transformational change (Kania et al., 2018). Structural change addresses areas such as policies, processes and resource flows. Described as an 'explicit' form of change, structural change is the most visible, with its results often readily observable and able to be assessed via traditional evaluation and measurement techniques (Kania et al., 2018). Structural change has tended to be the focus of efforts in Aotearoa, with approaches to addressing inequity typically following what Blanchard and Mukerjee (2022) refer to as a 'normative and technical' approach which assumes behavioural and attitudinal change will result from a combination of technical skills and competencies, the imposition of rules regulating resolution process and ethics, and oversight regulatory bodies. However, as Kania et al. (2018) conclude, whilst effecting structural change is important, it will not, without relational and transformative change targeting relationships and mental models, result in sustainable systems change. Supporting this, solutions premised upon technically oriented 'competency' approaches were recognised in Aotearoa over 20 years ago as ineffective, with systems operating under these labels continuing to privilege ways of thinking which delivered inequitable outcomes for Māori (Love, 2003).

Relational Change: Shifting Power

Relational change, a 'semi-explicit' condition of change, understands that system change essentially relies on transforming the relationships between people who make up the system (Kania et al., 2018). Relational change therefore focuses on the quality of connections, prioritising issues such as power and control, particularly for those who have been accorded limited power and decision-making authority. Genuine relational change often requires power to be surrendered at some level (Kania et al., 2018).

In Aotearoa, 'Kaupapa Māori', which uses Indigenous theorising as a vehicle for conscientisation, resistance and praxis, has over the past four decades actively disrupted the status quo, particularly in education and research spaces (Smith, 2017). Integral to Kaupapa Māori is the *principle of socio-economic mediation* which prioritises critically analysing and understanding Western knowledge bases and power structures to expose underlying assumptions that serve to conceal, perpetuate and maintain inequity for Māori (Smith, 2017). To not undertake this analysis sees contemporary forms of inequity isolated from the underlying structural colonially imposed power arrangements that dictate the relationship between Indigenous peoples and the State (Axelsson et al., 2016). Inequitable health outcomes for Māori are identified as being consistent with wider Indigenous experiences of colonisation which are characterised by land and language theft, and racist and discriminatory policies and actions (Graham & Masters-Awatere, 2020). Despite this, explanations which understand colonisation and colonialism as processes which continue to negatively impact health outcomes for Indigenous peoples have tended to be noticeably absent within the research base (Axelsson et al., 2016). Smith (2017) also emphasises the importance of becoming critically conscientised to the fact that colonisation can, and is perpetrated by, and against us, as Indigenous people. From a Kaupapa Māori theory perspective, as Axelsson et al. (2016) argue, colonisation must be brought into the analysis when considering Indigenous populations.

Consistent with Kaupapa Māori theory, Mahi a Atua has a central focus on the critique of power relations responsible for the deliberate and systemic marginalisation of mātauranga Māori (knowledge originating from a Māori worldview) and the resulting inequitable outcomes for Māori communities. The Mahi a Atua principle of *tēnei te pō, nau mai te ao* (Indigenising spaces), as a deliberate tool for systems transformation recognises that spaces, both physical and mental, have been dominated by a colonised world view which deliberately excludes and delegitimises Indigenous knowledge (Kopua et al., 2020). Indigenising one's space cannot occur without a critical lens on the dominant colonial paradigm. A central element of the institutional racism perpetuating inequity for Māori across health systems generally, and mental health systems specifically, is domination by a mono-cultural, illness-focused, bio-medical model in which 'expert' 'technical' knowledge is considered the primary source of authority (Kopua et al., 2020). Within this model, the experiences and expertise of those seeking assistance are accorded minimal validity and power, as they become positioned solely as passive recipients of the help provided by a technically proficient 'professional' (Bracken & Thomas, 2017).

Whānau as Change Agents

Building power and voice within communities is recognised as a powerful force for change, with systems change theory emphasising the importance of working together—relational change occurs when people and collectives across established boundaries and silos are engaged with and supported to work in ways not done before (Kania & Kramer, 2011; Senge et al., 2014). A seminal report in Aotearoa over 30 years ago, *Puao-Te-Ata-Tu*, recognised the community as central to system transformation (Māori Perspective Advisory Committee, 1988). Decades later, conclusions are still being drawn regarding genuine health system transformation in Aotearoa requiring all parts of the system to work differently and move beyond outdated professional boundaries and scopes of practice (Health & Disability System Review, 2020).

Communities positioned as 'champions of change' also continue to be considered the biggest untapped resource for systems change (Russell et al., 2018). However, from a systems change perspective, communities are more than just a 'resource'; they are *the* change-makers. As one systems theorist comments, "plugging the voice of the community into the right kind of political power grid will do more to create health and wellness than any other single intervention" (The California Endowment, 2016, p. 4). Fundamental to Kaupapa Māori theory, truly transformative Indigenous theories are those which are 'owned' and 'make sense' to those communities (Smith, 2003).

In Aotearoa, Whānau Ora, the paradigm which recognises that the wellbeing of individuals is indivisibly linked to the wellbeing of the collective (Taskforce on Whānau-Centred Initiatives, 2009), is one such transformative Indigenous theory. Acting as an overarching philosophy; a process and model of practice for service delivery; and a desired outcome goal (Boulton et al., 2013), Whānau Ora provides the foundation for an approach focused on flourishing whānau potential, as opposed to crisis and intervention (Smith et al., 2019). Whānau Ora recognises that even in times of distress, strengths and aspirations will always be present within whānau. Whānau growth, development and leadership is created when whānau are supported to tell their own stories, and activate their own resources and strengths (Kopua et al., 2021).

Consistent with both Kaupapa Māori theory and Whānau Ora, Te Kurahuna works from the ground up, investing in the power of communities and creating collective leadership in the form of a community of Mataora. Positioned across Aotearoa, Mataora are 'change agents' trained in Mahi a Atua, who influence and embed sustainable transformative change. Change agents' do not only exist within professional workforces, but across communities and whānau: anyone, both Indigenous and non-Indigenous has the potential to contribute to transformative change (Rangihuna et al., 2018b). As emphasised in Kaupapa Māori theory, it is not only those participating at policy and service levels who become conscientised in relation to the structural and systemic barriers for Māori, but more importantly it is whānau and communities at the flax-roots (Smith, 2005). Fundamental to a critical Kaupapa Māori theory is that these communities, through being provided with approaches by which to voice their ideas and aspirations, are supported to act transformatively via their everyday, individual and collective actions (Smith, 2003).

The Mataora workforce currently comprises over 2500 people. It includes the regulated clinical health workforce (e.g. general practitioners, psychiatrists, psychologists, counsellors, nurses, social workers, occupational therapists and midwives); the non-regulated health workforce (e.g. cultural advisors, whānau ora workers, community and peer support workers); managers, policy, decision-makers/commissioning staff, administrators, educators; Artists (tā moko, taonga pūoro, kiritao, musicians, writers, mau rākau, painters, actors, dance, kapa haka); and whānau members.

Transforming Mental Models

Although an essential element, the integration of mātauranga Māori in isolation will not lead to systems change (Smith, 2005). 'Transformative change', the final condition of systems change described by Kania et al. (2018) recognises that systems are formed by people, and it is the 'mental models' of those people, that is their deeply held beliefs, assumptions, frames of reference, and taken-for-granted ways of doing things, which determine their actions within those systems. These mental models create an implicit or unspoken bias through which everything that is thought, said and done is filtered (Kania et al., 2018).

Institutional racism and the practices which sustain it are heavily embedded within societal and cultural norms, often resulting from deeply ingrained mental models relating to personal, social and cultural environments (Blanchard & Mukerjee, 2022). Emphasising the power of these mental models, Hakio and Mattelmäki (2019) refer to the notion as being "paralysed as instruments trapped within our current paradigms" (p. 3). For example, specifically in relation to health care, Russell et al. (2013) conclude that whilst economic and geographic barriers impacting on service access are relatively straightforward to recognise and solve, barriers originating from a fundamental disconnect between Māori worldviews of health and wellbeing, and the disease-oriented medical model are not. When desired systems change aims to address equity, attention must be focused on exposing the mental models which are obstructing structural and relational change, particularly necessary power shifts (Kania et al., 2018). Issues exist across every system level: racist narratives; power dynamics and resource flows that reinforce existing power structures and relationships; and public policy underpinning inequitable outcomes (Kania et al., 2018).

Public policy decisions and processes are not ideologically neutral, heavily influenced by the normative cultural expectations of those designing them (O'Sullivan, 2019). In Aotearoa, State policies, processes and services have maintained the status quo (Boulton et al., 2020). For example, the dominant belief of 'equality' positions Indigenous models and services as undesirable for central government, with prescriptive contracts, short contracting periods and onerous accountability requirements for Indigenous providers, all considered symptomatic of a deeper desire on the part of the state to maintain control over Indigenous development. (Lavoie et al., 2016; New Zealand Productivity Commission, 2015; Smith et al., 2019). Inequity is not considered within the broader context of colonisation, and fundamental Māori concepts and processes are not understood by decision makers, despite the availability of a robust evidence base (Boulton et al., 2020).

Reid and Robson (2007) refer to how persistent inequity for Māori has become 'normalised', an almost expected and accepted feature of our national landscape. Much of this normalisation is sustained by mental models based on individualised deficit theory, language and indicators which underpin the stereotype that inequity results solely from the individual failings of Māori (Waitangi Tribunal, 2019). The normalisation of this stereotype then serves to justify the continued existence of inequitable service delivery, in the process endorsing the presence of ongoing institutional racism (Keddell, 2018; Rua et al., 2019). Narrow, static, individualised cultural competency acquisition frameworks also serve to perpetuate deficit discourses by attributing responsibility for problems to individuals and promoting over-simplified understandings based on cultural stereotypes, whilst simultaneously ignoring the systemic drivers of inequity, including the role played by health professionals themselves in creating and maintaining these inequities (Curtis et al., 2019). The combination of these mental models results in short-term reactive actions which are focused on producing quick results with the least risk (Kania et al., 2018).

The fundamental importance of transforming mental models is evident in the resistance we have encountered when attempting to operationalise Mahi a Atua, with much of the desire to maintain the status quo stemming from the self-preservation and self-interest of key system players (Kopua et al., 2021). Despite the importance of 'decentring' the expert and shifting power to the community being identified for over three decades now (see Māori Perspective Advisory Committee, 1988), these concepts remain challenging for those holding the power. Indigenous researcher and activist, Tina Ngata emphasises how the existence of these challenges highlights the difficulty of addressing issues of power within a deliberately created colonial system, as well as the complexities which present when people are being asked to dismantle the very system that is conferring the privilege upon them (Tina Ngata, Ngāti Porou, Aotearoa, personal communication, 1 October, 2021).

Creating New Ways of Thinking, Acting and Being

Understood as both the key to overall systems change, and the most challenging condition of systems change to address, Kania et al. (2018) argue that failing to shift and transform mental models means any structural and relational changes will only ever be incomplete and temporary. Unlike structural change, shifts in mental models are less visible and more difficult to measure, often requiring attention to 'the detailed and often mundane work of noticing and acting on much that is implicit and invisible to many' (Kania et al., 2018, p. 18). Real change starts with recognising that everyone is part of the system we seek to change (Kania et al., 2018). Summarising the work of Otto Scharmer and Katrin Kaufer, Senge et al. (2014) identify how well-intentioned systems change efforts fail because people are unable or unwilling to embrace the notion that change relies on opening the mind to challenge assumptions; opening the heart to vulnerability and truly hearing one another; and opening the will to let go of preset agendas and genuinely see not only what is really needed, but what is possible (p. 29).

Voluntarily relinquishing power within a system is a difficult, challenging and long-term process. Hakio and Mattelmäki (2019) stress the importance of 'pushing the boundaries' in relation to how we approach and solve problems. Integral to Mahi a Atua, where walking in the footsteps of our ancestors is not a rigid prescription, is that the human process of systems change will not be successful if it is based on rigid assumptions and short-term agendas. Kaupapa Māori theory, as a vehicle for conscientisation, resistance and praxis, emphasises the importance of 'reflection *and* action upon the world in order to transform it' (Freire, 1970, p. 51; emphasis added). As a 'way of being', Mahi a Atua facilitates movement to critical consciousness; that is, transformed mental models on which new actions can be based, with this shift occurring via an ongoing process of critical self-reflection and assessment; and examining structural variables such as power, social justice and equity, as well as one's own privilege and bias. Often personally confronting, particularly for those trained within dominant medical paradigms which serve to support and sustain ingrained systemic racism across health systems, this process explicitly includes one's own contribution to institutional racism (Kopua et al., 2021), that is, becoming critically conscientised to the fact that colonisation can, and is perpetrated by and against ourselves (Smith, 2017).

Creating new ways of thinking, acting and being requires deliberate and intentional processes of deep exploration, reflection and learning which illuminates taken-for-granted assumptions and bias, as well as the limitations of our existing mental models, and how they are connected to the problems we are seeking to solve (Senge et al., 2014). Transformative systems change prioritises reflective individual and collective processes and experiences which then provide the foundation for new patterns of thinking and behaviour (Blanchard & Mukerjee, 2022). Integral to an awareness-based systems-change model, is that changes in behaviours and new practices are not imposed, they emerge through personal inner transformations (Blanchard & Mukerjee, 2022).

Emerging programs that invite practitioners to engage in inner work are seen as responding to a growing understanding that complex systems thinking tools can be introduced in such a way that prompts a shift in one's modes of reasoning (Alexiuk, 2022). Cultivating the skills required to sit reflexively with challenging, unpleasant and paralysing, and often deeply buried feelings, bias and conflicts is a key part of this (Blanchard & Mukerjee, 2022; Hakio & Mattelmäki, 2019). Such work can only occur within candid and trustworthy environments, with facilitators experienced in holding space for groups as they embark on this type of transformative journey (Blanchard & Mukerjee, 2022).

Central to Mahi a Atua is the wānanga process, a taonga tuku iho (gift from the past), which intentionally creates an environment conducive to courageous and transparent collective learning, facilitating shifts in awareness and perspective, both individually and collectively. Mahi a Atua training asks Mataora to Indigenise their space—*Tēnei te Pō, Nau mai te Ao*: embedding Mahi a Atua principles across every element of their lives in order to rethink their usual way of being. With a focus on developing active learners who are responsive to whānau and the community, meaningful shared critical reflection also provides opportunities to try differing perspectives, enabling individuals and organisations to question, revise and release what Senge et al. (2014) refers to as embedded assumptions. Consistent with Kaupapa Māori theory which seeks to provide Māori communities with a voice by which to articulate both resistance and aspirations (Smith, 2017), an essential element of the paradigm shift sought by Te Kurahuna is that whānau awareness of the wider context in which their distress is positioned is deliberately enhanced.

The training environment and context provided by Te Kurahuna ensures Mahi a Atua does not simply replicate existing systems of competency acquisition. The Mahi a Atua principle of *ka mā te ariki, ka mā te tauira* privileges Indigenous ways of learning. With Mahi a Atua connected to whakapapa (genealogy), the direct relational connection between those who are teaching and those who are learning is understood. The process of wānanga in which Mataora are provided with approaches to understand and effect the paradigm shift necessary for structural, relational and transformative change is also utilised when working with whānau to develop meaningful responses to distress which privilege and reinstate the te ao Māori (Māori world) voice and culturally relevant meaning. Mahi a Atua wānanga emphasise āta whakarongo (active listening), kōrero (discussion), co-creativity and reflective communication in order to weave together the many points of whānau view about their distress and create new shared understandings in which everyone contributes important threads.

The Mahi a Atua principle of *Hongihongi te wheiwheiā* (inhale the unusual) focuses on letting go of pre-set agendas in order to see what is really needed, and in the process always striving to do better as an active learner. Integral to the wānanga process, including when working with whānau, is that reflective talk can assist in tolerating uncertainty; and when uncertainty is shared it can lead to being together differently. With outcomes for whānau prioritised and whānau positioned as the experts of their own experiences and aspirations, Mataora learn to be curious, imaginative and deliberate; active participants in a relational process where both parties are giving and receiving koha (gifts). Informed by Hinekauorohia, the Atua of healing and reflection, the importance of developing a culture of feedback, individually, collectively and organisationally is emphasised. Te Kurahuna Mataora are also continually monitoring for organisational and structural factors that facilitate or act as barriers towards achieving health equity for Māori.

Growing Collective Leadership and Imagination

Systems change occurs when system leaders focus on creating conditions that enable change to be self-sustaining (Senge et al., 2014). Aspirations for a collective critical consciousness and recognition of the collective power of individuals to effect systems change emerges from a uniquely Indigenous context. Challenging established concepts of leadership, Te Kurahuna works from the ground up, investing in the power of communities and creating collective leadership in the form of a critical mass of Mataora. Integral to the role of Te Kurahuna as a systems leader is the overt recognition that anyone has the potential to be an agent of change. The place and value of the many different and equally valuable contributions to systems change is upheld in this approach. For example, those facilitating alignment towards common goals and coordinating action across boundaries; those creating new solutions to address dysfunction; and those shining the light on system dysfunction wherever possible (Kimsey & Battilana, 2017) are all essential.

Effective systems change leaders will also move people away from seeing short-term reactive solutions as the answer to complex and challenging problems, shifting them instead to a process of collective imagining centred on realising opportunities for long term innovation (Senge et al., 2014). The systems change literature understands the importance of approaches by which people are able to articulate their deeper aspirations, alongside building confidence in what can be collectively achieved (Senge et al., 2014). System leaders are not separate from this process. Reflecting the Mahi a Atua principle of *ka mā te ariki, ka mā te tauira*, system leaders are always learners who will as Kania et al. (2018) identify, demonstrate a willingness to be vulnerable, rethink, reinvent and recommit. With the collective focus shifted from reactive problem solving to co-creating innovative futures, the tension between present day reality and the collective vision can serve to generate energy and momentum, attracting others to come on an aspirational journey (Senge et al., 2014).

The late Moana Jackson said, one of the biggest things taken away by colonisation was 'belief'; 'belief in the ability to make decisions for ourselves; belief that our language and culture was worth retaining; belief that our systems and ways of living were good enough' (Morrison, 2022). Utilising Freire's notion (1970), that the oppressed must also free themselves, Smith (2003) argues that the real revolution which occurred as a result of the development and application of Kaupapa Māori theory in the 1980s, was not so much progress made in relation to language revitalisation but in a 'reawakened Māori imagination'; the belief in our own ability to reflect and act in order to transform. Smith (2017) speaks about the mind-set transformations which occurred within Māori whānau and communities at this time. Decolonising discourses, via processes of critical consciousness, shifted to be significantly more proactive as the everyday decisions and actions of whānau, despite the presence of continued resistance, collectively created transformative change. Of importance, as Smith (2017) observes, 'every Māori is in the struggle whether they like it or not, and whether they know it or not' (p. 78).

Indigenous Systems Change: Healing the System

Rangi Parauri is the primary Mataora training programme offered by Te Kurahuna. Multi-modal and comprising both online and in-person learning, Rangi Parauri caters for up to 100 tauira (students) each intake and is delivered over a two-month period. Central to the learning process is the five-day in-person wānanga held on marae around Aotearoa, both rural and urban. Supported by detailed, daily online content, including pūrākau, Rangi Parauri is delivered by a team of highly skilled Mataora, including tohunga, te reo Māori exponents, moko artists, storytellers, weavers, mirimiri/romiromi practitioners, maramataka (Māori lunar calendar) practitioners, musicians, karetao (Māori puppetry), nurses, psychiatrists, psychologists and educators.

Using pūrākau and Indigenous approaches such as 'Taku Whare Wānanga' and maramataka, tauira are supported over the five-day wānanga to

- recognise when they are stuck and how they are contributing to the problem when they do not remain open to learning;
- understand their whakapapa connection, and how they are in relationship to the wider environment around them;
- explore the whakapapa of their emotions and healing; and
- develop their own strategies using Mahi a Atua principles, rituals and approaches.

Integral to the Rangi Parauri five-day wānanga is the availability of Tūporeariki for all tauira. Infused with pūrākau, maramataka, karakia and an Indigenous reflective rituals, Tūporeariki is a collective Indigenous wānanga healing space in which healing occurs via both personal and collective reflections. The final day of the Rangi Parauri ā-tinana wānanga has evolved into a celebration of the healing power of te ao Māori. With no pre-determined agenda, the final day naturally unfolds, whether it be people laying on the table to receive facial moko, or learning mōteatea under the trees outside.

Although Mataora work to bring healing to whānau in distress, Te Kurahuna also works to, via a paradigm shift which prioritises Indigenous knowledge as it directly confronts biomedical dominance, bring healing to the system itself. An aspect often overlooked in systems change theory, but of critical importance in a colonised context, is how the process of transformative systems change can bring healing to the system itself (Tina Ngata, Ngāti Porou, Aotearoa, personal communication, 1 October, 2021). There is evidence showing that as Mataora are empowered to work in ways which feel 'normal' for them, they become liberated in their practice (Tipene-Leach et al., 2019). Feedback recently received from health workforce tauira undertaking Rangi Parauri training demonstrates how Rangi Parauri wānanga provide opportunities for tauira to pause, reflect and heal within te ao Māori. It is becoming increasingly clear that by better understanding how colonisation has impacted connections to te ao Māori and mātauranga Māori, Rangi Parauri nourishes tauira as they are supported to see their worth and possibilities for healing. Although journeys for tauira are unique, integral to many is that they involve identity connection, reclamation, freedom and liberation. This healing and growth ripples out to whānau as tauira prioritise applying their learnings with their own whānau first, knowing that this is how they are able to activate Mahi a Atua as 'a way of being' across their entire lives.

Individual, intergenerational and collective mamae also exist as powerful forces across our systems today, impacting both on how the present moment is experienced, as well as the future course of systems. Unrecognised mamae not only prevent us from addressing the roots of the collective barriers we face, it prevents us from taking necessary steps towards healing that will transform the systems around us (Calderon de la Barca et al., 2024). That the 'healing' element of Rangi Parauri has emerged so strongly for tauira is indicative of the immense stress faced by the Indigenous health workforce in Aotearoa who experience institutionally racist systems on a daily basis.

Looking Forward

We have found that even after undertaking Mahi a Atua training, the pervasiveness of the dominant bio-medical paradigm sees issues such as the ongoing prioritisation of Western dominated approaches; difficulty adjusting to a transparent and feedback-informed practice; and being unwilling to consider issues regarding racism, implicit bias and inequity continuing to be present for individuals and organisations (Kopua et al., 2021). Tina Ngata highlights that despite genuine insight into the urgent need for transformative solutions, a tendency for people to rebound back to the status quo remains, with this largely due to the constant reaffirmation and replication of the colonial norm (Tina Ngata, Ngāti Porou, Aotearoa, personal communication, 1 October, 2021). That this is the case supports the view that transformative mental model change is not a linear, short-term process. Transformative change can occur only through sustained engagement over a long period of time (Blanchard & Mukerjee, 2022). As is detailed in systems change literature, transformative change requires developing new capabilities where much of the work occurs internally. This in turn necessitates long term commitment, patience, mentoring and ongoing regular practice (Kania et al., 2018; Senge et al., 2014). It is only by engaging in a long-term process that systems obscured by mental models are able to be fully revealed; engagement, trust and relationships gradually built; and the collective focus effectively shifted from reactive problem solving to one of emerging possibilities, innovation and creativity (Senge et al., 2014).

The long-term nature of transformative mental model change highlights the importance of access to continuing opportunities for reflection and generative conversations. Viewing mental model change as an ongoing process is fundamental, with Mahi a Atua deliberately positioned as a 'way of being' encompassing every aspect of one's personal and professional life. We are acutely aware that transitioning to a fully Indigenised equitable system requires a constant focus on the development of leadership able to champion ongoing skill development and fidelity with Mahi a Atua principles, practices and processes. Independence from dominant institutions and a commitment to ensuring attention remains focused on addressing institutional racism, particularly when challenging organisations at a strategic level, is integral to ensuring the integrity of Mahi a Atua is upheld and we remain uncompromisingly Indigenous.

Although Mahi a Atua was originally focused on mental health settings, there are growing numbers of Māori and non-Māori across wider health, education, justice and other social service sectors who have undertaken our training. This supports a documented growing demand for a far-reaching Mataora workforce able to effect systems change via a mātauranga Māori, whānau ora and whānau-centred knowledge and practice base (OECD, 2018). Because mātauranga Māori contains foundational principles and values that pertain to life in general, Mahi a Atua can be applied across any context in need of a collective shift in consciousness. Although sectors may differ, it is the same systemic issues which create, reinforce and sustain inequity for Māori. Recent observations made in the law profession highlight this, with Mahi a Atua considered able to support all lawyers, both Māori and non-Māori, to create the best legal system for Aotearoa (New Zealand Law Society, 2021). Mahi a Atua shifts the focus away from determining the quickest and least messy solution to legal issues, to seeing and understanding those within the justice system as products of colonisation, thus providing a platform from which to consider different solutions (New Zealand Law Society, 2021). Reflecting that Mahi a Atua is 'a way of being', systems can be changed at the same time whānau are being helped (New Zealand Law Society, 2021).

As was previously highlighted, addressing the institutional racism underpinning inequity for Māori is not simply about the acquisition of technical skills or competencies; it forms part of a wider liberation movement built on our own Indigenous methods and mechanisms (Baker & Levy, 2013). The depth and breadth of the transformative potential of Mahi a Atua for systems change is yet to be fully realised. This chapter has started to explore the linkages across knowledge bases in relation to the theoretical underpinnings of Mahi a Atua. Continuing to grow our evidence base is essential, particularly in relation to articulating and demonstrating how the knowledge, practice and processes of Mahi a Atua work to create a collective consciousness which harnesses the power of communities as change agents and provides a pathway to genuine and sustainable systems transformation.

Understanding transformative change itself requires a shift in the dominant mental models of assessment and measurement. The evidence supports the reality that genuine transformation is not a short-term endeavour; there are no shortcuts in the challenging critical self-reflection necessary for movement to critical consciousness, and the embedding of Mahi a Atua as a 'way of being'. Transformative change therefore occurs against a backdrop of constantly fluctuating activity, essentially making it impossible to determine 'cause and effect' in the way traditional linear evaluation or research frameworks typical demand (Kania et al., 2018, p. 15). Data able to reveal insights, often from within complex and changing reflective inner processes, needs to be gathered from multiple perspectives, utilising a range of methodologies, including those which are of priority for whānau (Hakio & Mattelmäki, 2019; Kania et al., 2018). Constantly exploring, improving and co-creating alongside our whānau who are engaging in Mahi a Atua alongside us is essential.

Te Kurahuna is a systems change entrepreneur. Viewed through an Indigenous lens, we have proximity to the communities and issues with whom we work; we engage with all those able to contribute skills, talents and expertise when developing innovative solutions; we view and embody collective leadership as a critical component of systems change; we focus on shifting multiple interrelated system conditions; we understand relationships and power dynamics; and we prioritise influencing transformative levers of systems change such as mental models and relational change, in addition to structural levers such as policy and resource flow. Added to this is the systems healing we integrate within our Indigenous paradigm of systems change. It is our capacity to activate a uniquely informed Indigenous systems change paradigm which propels Te Kurahuna far beyond that of being advocates simply calling for equity, to a powerful movement who are actively operationalising the key conditions of systems change.

We conclude this chapter with the reflections of Indigenous writer Mataora and clinical psychologist Lisa Cherrington (2022). Lisa asks:

Imagine if.
Instead of a court hearing we had a Mahi a Atua wānanga.

Where we held space for this young rangatahi, and the young person's whānau.
And the victim, and the victim's whānau.
And we were all tauira for a moment in time.
Where we also held space for the 'professionals' as well.
And we were all Mataora.
What if we were all agents of change,
activating our voice and speaking up to address racism?
What if we reflected how we too have played a part in the problem?
The judge. The lawyers. The psychologist. The psychiatrist.
What if we used Mahi a Atua principles and tikanga to hold this space?
Where we could all sit in te pō.
And through wānanga
and pūrākau
move into te ao.
Tēnei te pō, nau mai te ao.

References

Alexiuk, E. (2022). Telling sauna stories: A role for autoethnography in systems change research. *Journal of Awareness-Based Systems Change, 2*(1), 15–36. https://doi.org/10.47061/jabsc.v2i1.1976

Axelsson, P., Kukutai, T., & Kippen, R. (2016). The field of Indigenous health and the role of colonisation and history. *Journal of Population Research, 33*(1–7). https://doi.org/10.1007/s12546-016-9163-2

Baker, M., & Levy, M. (2013). E Toru Ngā Mea. *Pimatisiwin: A Journal of Aboriginal and Indigenous Community Health, 11*(13), 471–483.

Blanchard, V., & Mukerjee, D. (2022). *Awareness-based system change as the basis for transforming systems and social norms. UNDP Global Policy Network Brief.* https://www.undp.org/publications/dfs-awareness-based-system-change-basis-transforming-systems-and-social-norms

Boulton, A., Levy, M., & Cvitanovic, L. (2020). Beyond Puao-Te-Ata-Tu: Realising the promise of a new day. *Te Arotahi Paper Series, December* (Number 06).

Boulton, A., Tamehana, J., & Brannelly, P. M. (2013). Whānau-centred health and social service delivery in New Zealand: The challenges to, and opportunities for, innovation. *MAI Journal: A New Zealand Journal of Indigenous Scholarship, 2*(1), 8–32.

Bracken, P., & Thomas, P. (2017). Reflections on critical psychiatry. In B. Cohen (Ed.), *Routledge international handbook of critical mental health* (pp. 98–106). Routledge.

Brown, H., & Bryder, L. (2022). Universal healthcare for all? Māori health inequalities in Aotearoa New Zealand 1975–2000. *Social Science & Medicine.* https://doi.org/10.1016/j.socscimed.2022.115315

Calderon de la Barca, L., Milligan, K., & Kania, J. (2024). Healing systems. *Stanford Social Innovation Review.* https://ssir.org/articles/entry/healing-trauma-systems

Cherrington, L. (2022). Sad joke in clinical psychology. *Strength of Story: He Pūmanawa Kōrero.* https://lisacherrington.substack.com

Curtis, E., Jones, R., Tipene-Leach, D., Walker, C., Loring, B., Paine, S., & Reid, P. (2019). Why cultural safety rather than cultural competency is required to achieve health equity: A literature review and recommended definition. *International Journal for Equity in Health, 18*(174). https://doi.org/10.1186/s12939-019-1082-3

Freire, P. (1970). *Pedagogy of the oppressed.* Seabury Press.

Graham, R., & Masters-Awatere, B. (2020). Experiences of Māori of Aotearoa New Zealand's public health system: A systematic review of two decades of published qualitative research. *Australian and New Zealand Journal of Public Health, 44*, 193–200. https://doi.org/10.1111/1753-6405.12971

Hakio, K., & Mattelmäki, T. (2019). Future skills of design for sustainability: An awareness-based co-creation approach. *Sustainability, 11*(19), 52–47. https://doi.org/10.3390/su11195247

Health & Disability System Review. (2020). Health and disability system review – Final report – Pūrongo Whakamutunga. https://systemreview.health.govt.nz/

Health Quality & Safety Commission. (2019). A Window on the Quality of Aotearoa New Zealand's Health Care 2019. Retrieved from https://www.hqsc.govt.nz/our-programmes/health-quality-evaluation/publications-and-resources/publication/3721/

Kania, J., & Kramer, M. (2011). Collective impact. *Stanford Social Innovation Review, Winter*, 36–41.

Kania, J., Kramer, M., & Senge, P. (2018). The water of systems change. https://policycommons.net/artifacts/1847266/the-water-of-systems-change/2593518/

Keddell, E. (2018). The vulnerable child in neoliberal contexts: The construction of children in the Aotearoa New Zealand child protection reforms. *Childhood, 25*(1), 93–108. https://doi.org/10.1177/0907568217727591

Kimsey, M., & Battilana, J. (2017). Should you agitate, innovate, or orchestrate? *Stanford Social Innovation Review*. https://doi.org/10.48558/3YGB-3M56

Kopua, D., Kopua, M., & Bracken, P. (2020). Mahi a Atua: A Māori approach to mental health. *Transcultural Psychiatry, 57*(2), 375–383. https://doi.org/10.1177/1363461519851606

Kopua, D., Kopua, M., & Levy, M. (2021). *Te Whare Wānanga o Te Kurahuna: Tēnei te Po Nau Mai Te Ao – Transformation in Action*. https://www.mahiaatua.com/hauakaaka/

Lavoie, J., Kornelsen, D., Wylie, L., Mignone, J., Dwyer, J., Boyer, Y., Boulton, A., & O'Donnell, K. (2016). Responding to health inequities: Indigenous health system innovations. *Global Health, Epidemiology and Genomics, 1*(e14).

Love, C. (2003). Keynote address. In L. Nikora, M. Levy, B. Masters, W. Waitoki, N. Te Awekotuku, & R. Etheredge (Eds.), *The proceedings of the national Māori psychology graduates' symposium 2002: Making a difference*. Māori and Psychology Research Unit, Department of Psychology, University of Waikato.

Māori Perspective Advisory Committee. (1988). *Puao-te- Ata-tu (day break): The report of the Ministerial Advisory Committee on a Māori Perspective for the Department of Social Welfare*. Department of Social Welfare.

Morrison, S. (2022, April 2). [@formelydaniels]. Neph @HynesSafari had this conversation with Dr Te Moana nui a Kiwa (Moana) Jackson. Another # MoanaMicdrop #believeinMāori. Image Attached] [Tweet]. Twitter. https://twitter.com/formerlydaniels/status/1509990526359007234

New Zealand Law Society. (2021). Mahi a Atua – Using stories to fulfil big dreams of systemic change. *Kōrero mō Te Ture – LawTalk, 947*, 9–13. https://www.lawsociety.org.nz/news/publications/lawtalk/lawtalk-issue-947/

New Zealand Productivity Commission. (2015). *More effective social services*. https://www.productivity.govt.nz/

O'Sullivan, D. (2019). *Waitangi Tribunal: Wai 2823 – Brief of evidence*. Waitangi Tribunal.

OECD. (2018). Mental health and work: New Zealand. https://www.oecd.org/newzealand/mental-health-and-work-new-zealand-9789264307315-en.htm

Rangihuna, D., Kopua, M., & Tipene-Leach, D. (2018a). Mahi a Atua: A pathway forward for Māori mental health? *New Zealand Medical Journal, 131*(1471).

Rangihuna, D., Kopua, M., & Tipene-Leach, D. (2018b). Te Mahi a Atua. *Journal of Primary Health Care, 10*(1), 16–17. https://doi.org/10.1071/hc17076

Reid, P., Cormack, D., & Paine, S. (2019). Colonial histories, racism and health – The experience of Māori and Indigenous peoples. *Public Health, 172,* 119–124. https://doi.org/10.1016/j.puhe.2019.03.027

Reid, P., Paine, S.-J., Te Ao, B., Willing, E., Wyeth, E., & Vaithianathan, & Loring, B. (2022). Estimating the economic costs of Indigenous health inequities in New Zealand: A retrospective cohort analysis. *BMJ Open, 12.* https://doi.org/10.1136/bmjopen-2022-065430

Reid, P., & Robson, B. (2007). Understanding Health Inequities. In B. Robson & R. Harris (Eds.), *Hauora:Māori Standards of Health IV – A study of the years 2000–2005* (pp. 3–10). Te Rōpū Rangahau Hauora a Eru Pōmare.

Reid, P., Robson, B., & Jones, C. P. (2000). Disparities in health: Common myths and uncommon truths. *Pacific Health Dialog, 7*(1), 36–47. http://pacifichealthdialog.org.fj/Volume207/No120Maori20Health20in20New20Zealand/Original20Papers/Disparities20in20health20common20myths20and20uncommon20truths.pdf

Rua, M., Hodgetts, D., Stolte, O., Cochrane, B., Karapu, R., Neha , E., Chamberlain, K., Te Whetu, T., Te Awekotuku, N., Harr, J., & Groot, S. (2019). Precariat Māori Households Today. *Te Arotahi Paper Series.* http://www.maramatanga.co.nz/node/1153

Russell, L., Levy, M., & Cherrington, L. (2018). *Whakamanawa. Honouring the voices and stories of Māori who submitted to the 2018 Government Inquiry into Mental Health and Addiction in Aotearoa. Unpublished Report.*

Russell, L., Smiler, K., & Stace, H. (2013). Improving Māori health and reducing inequalities between Māori and Non-Māori: Has the Primary Health Care Strategy Worked for Māori? https://www.wgtn.ac.nz/health/centres/health-services-research-centre/our-publications-archived/reports/phcse-maori-report.pdf

Senge, P., Hamilton, H., & Kania, J. (2014). The dawn of system leadership. *Stanford Social Innovation Review, 13*(1), 27–33. https://ssir.org/articles/entry/the_dawn_of_system_leadership#

Smith, G. (2003). *Kaupapa Māori theory: Theorizing indigenous transformation of education and schooling.* Joint AARE/NZARE Conference, Auckland.

Smith, G. (2005). Beyond political literacy: From conscientization to transformative praxis. *Counterpoints, 275,* 29–42. https://doi.org/http://www.jstor.org/stable/42978775

Smith, G. (2017). Kaupapa Māori Theory: Indigenous transforming of education. In T. K. Hoskins & A. Jones (Eds.), *Critical conversations in Kaupapa Māori* (pp. 71–81). Huia Ltd.

Smith, V., Moore, C., Cumming, J., & Boulton, A. (2019). Whānau Ora: An indigenous policy success story In J. Luetjens, M. Mintrom, & P. t. Hart (Eds.), *Successful public policy: Lessons from Australia and New Zealand*. ANU Press. https://doi.org/10.22459/SPP

Taskforce on Whanau-Centred Initiatives. (2009). Whanau Ora: Report of the Taskforce on Whanau-Centred Initiatives. Retrieved from https://www.msd.govt.nz/documents/about-msd-and-our-work/publications-resources/planning-strategy/whanau-ora/whanau-ora-taskforce-report.pdf

The California Endowment. (2016). *A new power grid, building healthy communities at year 5*. http://www.calendow.org/bhcreport/

Tipene-Leach, D., Able, S., Hiha, A., & Matthews, K. (2019). *Rangahaua Te Kūwatawata Final Report*. Māori & Indigenous Research, Eastern Institute of Technology.

Waitangi Tribunal. (2019). *Hauora: Report on stage one of the health services and outcomes Kaupapa Inquiry (WAI 2575 Waitangi Tribunal Report)*. https://forms.justice.govt.nz/search/Documents/WT/wt_DOC_152801817/Hauora%20W.pdf

Part III

Self-Surveillance, Contagion and Pathologisation

8

Digitalising the Vagus Nerve: Critical Mental Health and the 'total body'

Jacinthe Flore

Introduction

The digitalisation of mental health is an expansive phenomenon. It includes devices such as smartphone applications (apps) and wearable devices (wearables), as well as digital content related to mental health on websites and social media. Where the latter are often more focused on providing advice, offering positive encouragement, and detailing experiences, apps and wearables can mine and interpret data from the body to provide personalised mental health advice to users. Digitalisation has also transformed professional practice, both through well-known technologies such as teletherapy or telepsychiatry (May et al., 2001; Zeavin, 2021), the monitoring of medication intake using ingestible sensors (Flore, 2021), and the more recent emergence of generative artificial intelligence (Cheng et al., 2023). These examples demonstrate the

J. Flore (✉)
The University of Melbourne, Melbourne, VIC, Australia
e-mail: jacinthe.flore@unimelb.edu.au

N. A. Hendry, E. Karageorgos (eds.), *Critical Mental Health in Australia and Aotearoa New Zealand*, The Politics of Mental Health and Illness,
https://doi.org/10.1007/978-3-031-82861-4_8

breadth of developments in digital mental health. Not only are these technologies changing how mental health and ill health are experienced (Flore, 2024), but they are also transforming what is meant by 'mental health' and 'mental ill health'.

Wearable devices, in particular, collect and use data beyond the traditional areas associated with mental health (i.e. the brain, the mind, and/or emotions). Rather, the entire body as well as its socialisation have become important for capturing and addressing mental health and ill health. Hence, mental health is 'everywhere' in the body. By encouraging certain practices of data collection to achieve mental wellness, wearables build an embodied harmony as significant, as healthy, and as requiring work. This chapter seeks to explore how mental health becomes a matter for what I will call the 'total body', that is mental health that exists and requires labour beyond the brain and encompasses the body in its totality. The chapter uses as a case study the digitalisation of the vagus nerve through wearables.

The vagus nerve, as I will discuss, has been positioned as a uniquely potent producer and regulator of mental health in recent years. Through the digitalisation of the vagus nerve, mental health is produced as distinctively relational; one's attunement to the body is the focus of these technologies. This chapter will first contextualise the emergence of the vagus nerve as a site of and for mental health in relation to the broader medical and popular interest in the microbiome-gut-brain axis. It then discusses some key examples of the digitalisation of the vagus nerve with reference to wearable devices that aim to act on the nerves and produce wellness. The chapter finally draws on ideas of sociomaterialism to discuss the particular kind of bodily attunement that is produced at the vagus nerve, mental health, and digitalisation nexus.

Vagus Nerve Stimulation: A Brief History

From the Latin *vagus*, meaning wanderer or vagrant, the vagus nerve—also known as the vagal nerve—is the tenth cranial nerve. It is an extensive network of nerves inside the human body. The vagus nerve is a pair of nerves that acts as a key connector of brain, heart, lungs, and abdominal organs, and it is the main component of the parasympathetic nervous system (Faller & Schuenke, 2004, p. 163), as well as a central part of the microbiome-gut-brain axis. In fact, the vagus nerve is the main neurological pathway of the microbiome-gut-brain axis. In the clinical research on the connection between the axis and the nerve, the vagus nerve is described as able to "sense gut bacteria and their metabolites and transfer this information to the central nervous system" (Mörkl et al., 2023, p. 2). The uniqueness of the vagus nerve lies in both its expansiveness and multiple coils in the body, and its capabilities, as it is understood to play a wide role in health, including gut health, wellbeing, relaxation, and emotions (Mörkl et al., 2023). Its integration with the microbiome-gut-brain axis has led to widespread popular and cultural interest, and in 2023, several articles were published online assessing the potentials of vagus nerve and its role in the microbiome-gut-brain axis (see, e.g., Davis, 2023; Frankel & Warren, 2023; Larkin, 2023).

The vast presence of the vagus nerve in the human body makes it a complex, wandering nerve, a nerve that both journeys in the body and is responsible for ferrying essential messages throughout the body. It is also, perhaps, a nerve that is unique for having been shrouded in mystery until the late 1990s (see Borovikova et al., 2000) despite, as we will see, the documentation of its potentials since the nineteenth century. The vagus nerve has been the subject of fascination for centuries. The production of discourses on this nerve extends from scores of self-help books that promise to help readers 'activate', 'power', 'unleash', and 'soothe' the vagus nerve,[1] to wellness accounts on social media that coach users through exercises and movements that can reset and balance the vagus nerve, and online media articles on the possibilities of the nerve. More broadly, the role of nerves in shaping health and illness features prominently in historical texts in medicine and psychiatry. This history is outside the scope of this chapter, but it is worth highlighting that the late-eighteenth and nineteenth centuries in western European countries and in North

[1] Examples of titles found on online bookstores include: *Accessing the healing power of the vagus nerve: Self-help exercises for anxiety, depression, trauma, and autism*; *Activate your vagus nerve: Unleash your body's natural ability to heal*; *Anchored: How to befriend your nervous system using polyvagal theory*; *The vagus nerve gut brain body connection: Heal your vagus nerve and improve gut health*; and *Vagus nerve: Activate the secret power source within your body.*

America witnessed the expansion of multiple nervous disorders including hysteria and neurasthenia. For example, in the nineteenth century, the French neurologist Jean-Martin Charcot viewed hysteria as an inherited disease of the nerves (Charcot, 1875), while the American George Miller Beard, also a neurologist, cautioned against an epidemic of neurasthenia (weak nerves), for which urban dwellers were particularly at risk (Beard, 1881).

Medical interest in the nervous system spans the *longue durée* of the nineteenth century in several western countries. Likewise, the vagus nerve and its association with health and illness have a lengthy history. For the purposes of this chapter, it is important to note the contribution of James Leonard Corning, an American neurologist, who is known for pioneering spinal anaesthesia (Trent, 1946). In the 1880s, Corning designed multiple instruments, such as the carotid fork and the electrocompressor, that could stimulate the vagus nerve using electricity. This technique would affect different bodily functions, including an individual's heart rate and cerebral blood flow. His tools stand as early examples of vagus nerve stimulation. At the time, Corning's research was focused on the treatment of epilepsy, though he wrote on the "marked improvement in general health" of his patients (Corning, 1882, p. 173). However, Corning's claims of successful treatment were dismissed as exaggerated, and his methods and instruments were mostly shunned by his contemporaries.

It was not until a century later that vagus nerve stimulation was revived in medicine, and clinical researchers and pharmaceutical corporations began to explore the vagus nerve in earnest. In the 1990s, following animal experiments, the anticonvulsive potential of vagus nerve stimulation was 'rediscovered' (Moroşanu et al., 2023). While clinical research in vagus nerve stimulation in the 1990s mostly centred on the treatment of epilepsy, it was not long before mental ill health became an area of possible application. It is, of course, not possible to date precisely the turn to mental ill health. However, some indicators are useful. For example, in 2005 the Food and Drug Administration (FDA), a federal agency of the United States' Department of Health and Human Services, approved vagus nerve stimulation as an adjunctive treatment for the "long-term treatment of chronic or recurrent depression for patients 18 years or older

[...] and [who] had not had an adequate response to four or more adequate antidepressant treatments" (Food and Drug Administration, 2005). The system, named SYMMETRY and developed by the company LivaNova, is described as a tool to treat "difficult to treat depression" (LivaNova, 2023).[2]

According to the patient pamphlet, SYMMETRY is an implantable technology that requires surgery. It can be fitted inside the body, below the collar bone. It also includes a lead from the device that is connected to the left vagus nerve in a patient's neck. SYMMETRY stimulates the vagus nerve by "regularly sending mild pulses through a thin wire directly to the vagus nerve in the neck" (LivaNova, 2021, p. 3). As well as improving symptoms associated with depression, the system "shows significant improvement in quality of life", including mood, overall wellbeing, and sexual function (LivaNova, 2021, p. 6). Implanted vagus nerve stimulation is available in several countries, including Australia, the United Kingdom, and the United States.

From manual and electrical stimulation to twenty-first-century implantable technologies and digitalisation, the vagus nerve has generated enthusiasm among mental health researchers and technology developers alike. Besides SYMMETRY, which has been classified as a medical device by the FDA and other regulatory agencies, there are multiple devices boasting the ability to reset and stimulate the vagus nerve to nurture mental wellness. SYMMETRY's approval in 2005 can be said to have paved the way for other commercially available devices that do not require surgery and offer different experiences to users. Indeed, many of these technologies replicate well-known features of existing apps and wearables for mental health, for example, a focus on the design principles of UX or user experience in technology development, which includes enabling the user to tailor the device according to their individual needs—a function that SYMMETRY does not afford. SYMMETRY is also well

[2] 'Difficult to treat depression' is a recent category of depression that aims to reconceptualise 'treatment-resistant depression'. It is used to refer to individuals diagnosed with major depressive disorder who continue to experience significant distress despite traditional treatments, which can include psychotherapy and pharmaceutical drugs. Some psychiatrists claim that the nomenclature does not refer to a 'new' diagnosis but rather, "a framework or model of care" (McAllister-Williams et al., 2021). The utility of this concept remains contested (Cosgrove et al., 2021).

established as within the purview of medical expertise. However, as critical mental health scholars have argued, when we speak of digital mental health, 'expertise' is a fluid concept with porous edges (Crosby & Bonnington, 2020; Fullagar et al., 2017; Hendry, 2020). As Simone Fullagar and colleagues observe, the discourses circulated by mental health technologies do not merely impart expertise upon passive users; rather, they

> invite the subject to enfold and enact particular forms of expertise – visualising their problems in terms of thoughts and feelings, monitoring and actively acquiring new insight—to adjust their lifestyles, embodied habits and goals in the pursuit of optimised 'normality'. Signs of 'abnormality' (overwhelming affects, complex biographies and social injustices) figure as problems to be overcome and pleasurably mastered through learning new forms of expertise. (Fullagar et al., 2017, pp. 8–9)

Claims of knowledge about mental health exceed psychiatric or psychological research spaces and clinics, and technologies can both reaffirm and disrupt expert knowledge about mental health and ill health while also being embroiled in commercial interests that are dedicated to generating profit. These companies are also based in various areas, especially Europe, the United Kingdom, the United States, and they ship their products internationally. Hence, while SYMMETRY is located within the control of biomedicine, the devices discussed below harness and foreground connections to clinical sciences to produce commercially and internationally available technologies. In doing so, vagus nerve wearables transform knowledge and experience of mental health; they are an important example of the reshaping of epistemologies and ontologies of mental health.

Digitalising the Vagus Nerve

Wearable devices are a ubiquitous form of embodied computing or computational embodiment, which refers to technologies that circulate on, in, and around the body (Pedersen & Iliadis, 2020, p. xvii). Wearables for mental health are becoming increasingly accessible to users. One of the most popular wearables in the world, Fitbit, quickly transformed from an exercise and fitness wearable at its inception in 2009, to the filing of a patent in 2021 that indicates the company's interest in mental health, with mentions of depression and bipolar disorder in the document (Heneghan et al., 2021). As I have argued elsewhere, patents cannot be viewed as clear indicators of the intentions of a company[3]—often patent filing relates more to commercial monopolies than actual product development; however, patents reveal ideas and ideals, visions, and possible futures and profits that corporations imagine (Flore, 2023, p. 55). In this manner, patents gesture towards imaginaries of commercialisation, while also revealing in particular a conceptualisation of the body, the mind, mental health, and mental ill health.

This interest in mental health is, of course, not new. Much like the Apple smartwatch, Fitbit has long positioned its device as a tool of stress management that can improve mood. Terms such as 'mood', 'wellbeing', 'emotion', and 'stress' are often deployed on these wearables as proxies for mental health, and certain metrics are seen as correlations of mental states. These bodily metrics, in the case of FitBit, include heart rate, sweat levels, and skin temperature (Perez & Abdel-Ghaffar, 2023). While the connection to mental health might be accomplished using generalities or euphemisms—that is 'mood', 'wellbeing', 'emotion', and 'stress' instead of 'mental health'—for some clinical mental health researchers, wearables are promising. One of the reasons for this is their enabling of continuous and passive quantification of users' bodies without the need for individuals' verbal reports on how they feel whether through an interview or a questionnaire (e.g. Dunn et al., 2018). Thus, where words, stories, and narrative are viewed as subjective or even possibly embellished, data from the body are closer-to-infallible, and the body cannot lie about how the subject is feeling.

This represents an augmented digital phenotyping whereby the body as a whole must be mined for signs of stress and mental distress. For Isabel Pedersen, the continuous tracking of bodily signs and functions is

[3] Google, a subsidiary of Alphabet Inc., purchased Fitbit in 2019 and completed its acquisition of the company in 2021.

often "combined with the value system that assumes automated feedback is superior to human-interpreted information or human feelings" (Pedersen, 2020, p. 33). Here, human feelings as mapped onto the surface of bodies become data; data purportedly reflect feeling with no bias. Tracking moods during the day using emojis—enabled by apps, for example—is amplified through wearables. In fact, wearables typically are accompanied by apps that can help users personalise and track their experience of the technology. Increasingly, a new generation of wearable devices claim to draw on all human senses to offer therapeutic benefit through sounds, vibrations, and electrical stimulations; to illustrate, I turn to three examples below.

Sensory activation or stimulation is a central part of the work of wearables that offer vagus nerve stimulation or vagus nerve toning. One such example is Apollo, a wearable that can be worn on different parts of the body including the wrist, ankle, waistband, and on a bra strap or top. Apollo is among the most popular devices for vagus nerve stimulation on Amazon's online store. Apollo does not have a screen to act as a desktop or dashboard, but a connected app can be downloaded to a smartphone, smartwatch, or tablet. The company describes Apollo as a technology that "delivers soothing vibrations [...] that are like music your body can feel — higher vibrations help improve your energy and focus, while lower vibrations reduce stress and help you relax" (Apollo Neuro, 2024). The website for Apollo also emphasises that the technology was "born from neuroscience research" (Apollo Neuro, 2024). Users can use the app to create a personalised schedule of vibes "to play throughout your day and night to maximize your energy levels, sleep, and overall well-being" (Apollo Neuro, 2024). On its blog, the company draws connection between the device and vagus nerve stimulation. A blog post from 2023 explains the opportunities that Apollo sees in the vagus nerve. The blog explains under the sub-heading "Assist your vagus nerve in calming your body and mind with Apollo":

> As we uncover its potential in various fields, we find a parallel in the Apollo wearable – a device designed to support your bodily rhythms and create a balanced, harmonious state of being. (Apollo Neuro, 2023)

An earlier blog post from 2022 also describes the vagus nerve as the "body's information superhighway" and describes how the technology can help soothe the nervous system through the power of touch. The wearable's "vibrations mimic natural oscillation patterns between the heart and lungs activated during deep breathing" (Apollo team, 2022). Other wearables mirror a similar discourse wherein (1) the vagus nerve is described as bursting with opportunities for wellness, and (2) while several exercises *sans* technology can help activate or soothe the vagus nerve to promote health, wearable technologies can tap into these 'natural' functions and enhance them. Thus, the digitalisation of the vagus nerve amplifies benefits that can be achieved through certain exercises (e.g. controlled breathing at different speeds, exercises, activities such as yoga and singing, and meditation). What wearables afford, however, is a unique insight into the body's work and a way to understand and 'control' the nervous system to reduce stress and improve mental health.

Several other wearables are available on the market, and some degree of legitimacy is afforded to them when they are included in media articles. One such product is Nurosym, which was featured in an article in *The Guardian* (see Geddes, 2023). Nurosym is a product of the company Parasym. This wearable functions through an earpiece and a small remote control with different options for controlling electrical signals. The website explains:

> Nurosym harnesses advanced neuromodulation technology in a sleek and easy-to-use wearable device. Nurosym sends patented electrical signals to the brain via the vagus nerve, achieving specific therapeutic mechanisms that can counter dysregulation in the nervous system and significantly improve health outcomes, as proven by clinical studies. Nurosym is a certified medical device and the most studied non-invasive vagal neuromodulation system in the world. (Parasym, 2024b)

Electric signals can be transmitted to the brain through the route of the vagus nerve and thus 'counter dysregulation', meaning re-balance nerves that have become dysregulated. The deployment of 'signals' to the vagus nerve and the brain is reproduced by other wearables. This includes Sensate, a wearable which aims to "gently soothe your nervous system using non-invasive resonance on your sternum" (Sensate, 2024a) while its website shores up the importance of the vagus nerve and describes it as "the gut-brain superhighway" (Sensate, 2024b). Uniting these three examples is their material qualities as technologies and not pharmaceutical drugs. For example, Nurosym notes that the primary approach of "prescription pharmaceuticals" to mental health struggles "leaves many at a loss" (Parasym, 2024a). While claims that vibrations can assist with stress and mental health have been contested (e.g. Shmerling, 2021), these wearables continue to flourish, and they tend to underline how their approach is not invasive. The technology over drugs approach is presented as a more natural way of addressing ill health.[4] Several corporations emphasise the 'natural' aspects of their technologies; in particular, they claim to harness and fine-tune the already existing and natural functions of the body; in other words, nothing 'new' or 'foreign' is introduced.

Attuning to the 'total body'

Against this background, vagus nerve wearables offer an alternative to chemicals while also unifying the body, which I discuss below. Here, the body is not transformed by chemicals, but rather, technology can enable a sharper attunement to one's body. Natural (i.e. non-pharmaceutical or non-chemical) is equated with a sense of moral 'good'. This echoes Peter Conrad's observation that "the pursuit of wellness is inherently a moral pursuit, the achievement of a sense of virtue" (Conrad, 1994, p. 398). This quest assigns value to the body and privileges a medication-free existence. While Conrad's research was conducted before the emergence of digital wellness cultures and their wider reach, his assessment resonates with more recent comments about such cultures and their ability to

[4] The idea that 'data' or 'technology' are less invasive or problematic than pharmaceutical drugs has a social and political history that cannot be fully addressed here. However, it is important to note that this idea has gained more traction in recent years, particularly in the wake of mounting criticisms in mental health about 'overdiagnosis' and marketing practices of the pharmaceutical industry (Frances, 2014) and, more broadly, high-profile scandals such as the opioid crisis (see Trabsky & Flore, 2024).

produce 'wellness' as "a symbolic good, a marker of moral taste and a bodily discipline" (Hendry, 2023, p. 665).

And yet, bodily discipline through vagus nerve technology might require less work, less pursuit from the user. The technologies discussed in this chapter all target the nervous system in some form. However, where other wearables such as Fitbit have been examined as tools of self-knowledge and improvement, and self-surveillance, and necessitate commitments from users (see, e.g., Esmonde & Jette, 2020), the new generation of vagus nerve wearables do not require much effort of users, except of course to set time aside to use the device. This is a significant change from apps and wearables that require some form of user input and/or user interaction. Here, one can passively use the devices and achieve wellness with and through the nervous system. However, rather than congealing the location of mental health as matter of brain, mind, or, in these examples, nerves, the total body is involved in the production and experience of mental health. Sensate is placed on the sternum, Nurosym is placed on the ear, and Apollo may be positioned in multiple places on the body. What is shaped through vagus nerve stimulation and its purported ability to impact mental health positively is a kind of bodily attunement that, I argue below, transforms the epistemology and ontology of mental health.

In her influential book *Meeting the Universe Halfway* (2007), feminist theoretical physicist Karen Barad argues that we ought to think about ontology and epistemology as inextricable and proposes the term onto-epistemology. As Barad argues, "knowing is a material practice of engagement as part of the world in its differential becoming" (Barad, 2007, p. 89). Originally used in Barad's work alongside ethics to refer to how knowledge production is inseparable from practice, experience, and human and non-humans, this conceptualisation has proven fruitful in critical scholarship focused on the body, health, and mental health (e.g. Fullagar et al., 2019; Lupton, 2019; Pavlidis, 2023). This frame, which finds a productive footing within the fields of feminist new materialism and sociomaterialism, enables an understanding of mental health as complex, worldly, relational, and irrevocably embodied. For the rest of this section, I wish to draw on these ideas to discuss the bodily attunement that is produced at the vagus nerve, mental health, and digitalisation nexus.

Vagus nerve wearables, as noted above, invoke the language of harmony, balance, and regulation. They are positioned as technologies that can help users work towards an equilibrium within the body. This is achieved with and through nerves. However, these wearables go further in their conceptualisation of mental health. By encouraging users to evaluate moods, emotions, situations, and resting and exercise habits, they suggest an onto-epistemology of mental health that extends beyond the mind or the brain. Their primary work may be upon the nerves, but this does not mean a kind of localised nervous system fragmented from the body. The vagus nerve and its qualities as an expansive network—itself within another network, that is the microbiome-gut-brain axis—are drawn upon to situate mental health everywhere in the body. In other words, achieving mental health requires labour upon the total body, just as mental health is ontologically and epistemologically placed as circulating in (and around) the body. The positioning of mental health and ill health as a matter for the total body suggests that the relationship between human and technologies is not simply about data revealing states of wellness and illness. Rather, these states and experiences are co-produced and entangled. In this perspective, data are not 'just' representing the body; data are co-generative of mental health and ill health through mining and processing bodily signals.

This suggests an entangled system where the brain and the mind are *not* the primary generators of mental ill health or mental wellness. This argument builds, in particular, on the work of feminist scholars who have examined the role of the gut in producing mental health and breaking down and absorbing medication (see Dryden, 2023; Lucas, 2018; Wilson, 2015). For example, Grace Lucas (2018, p. 5), in a discussion of mental health and the microbiome-gut-brain axis, argues that "[t]he gut responds, conveys and unfolds mood. It is not merely a narrative representation or a metaphor or a secondary somatic response to cognitive thinking." In the case of the vagus nerve, it is the distinctive qualities of the nerve itself—expansive, entangled, and evolving—alongside its digitalisation that dislodges mental health from the mind and the brain and produces it as embodied. Mental health thus becomes achievable through attunement and relationality. This attunement requires labour to enhance the relationality to individuals' bodies and their social and cultural life-worlds. Crucially, this attunement can be generated and personalised with technologies.

Conclusion

Across the nineteenth to the twenty-first centuries, various models have emerged to inform biomedical approaches to mental ill health particularly in western countries. Most of these approaches have tended to focus on a fragmented body where mental ill health is shaped by and located in the mind or the brain, rather than considering mental health and ill health as embodied and relational. Neurological frames have also resulted understandings of mental ill health as related to serotonin and dopamine productions—the 'brain chemicals' that strongly influence (if not determine) one's mental state. These approaches have been heavily criticised, including by clinicians, both because of its severing of the mind from the body and for its connections to for-profit pharmaceutical industries, which further legitimatised theories of chemical imbalance (see, e.g., Moncrieff, 2008, 2013).

The search for the biological origins of mental ill health has preoccupied some areas of medicine for centuries. Unable to visualise mental ill health through a blood test or, for most diagnoses, a brain scan, behaviours, moods, and experiences are recorded over time and often interpreted as 'disorders'. The frame of mental ill health as neurological disorder—and hence a problem for the mind and the brain—continues to be one of the dominant approaches in clinical settings. This is despite the rise of biopsychosocial models, which have also been criticised as neglecting the affective and relational dynamics of mental ill health (see Fullagar et al., 2019). Mental health researchers and social activists have long called for a less medicalised, more 'holistic' and complex approaches that would account for social, economic, and cultural difference, as well as tackle stigma and marginalisation (e.g. Gorman & LeFrançois, 2017; Tseris et al., 2022). In some ways, digital mental health could be viewed as poised to address these thorny issues. By targeting mental health by using familiar, mundane concepts such as 'mood' and 'stress' on apps and wearables, mental health becomes an everyday experience,[5] an aspect of life that everyone must tend to, and something easily understood through data visualisation.

Hence, broadening of mental health beyond the brain and the mind could be viewed positively. It suggests that a less deterministic and more socially sensitive conceptualisation of mental health and mental ill health is possible. Indeed, vagus nerves wearables powered by advanced algorithms including artificial intelligence can offer unique, bespoke, and highly personalised experiences to users. The technology can be tailored to individuals and their specific social and cultural worlds. On the other hand, such an approach renders data collection essential, meaning that only large and continual mining of the body's responses by corporations can lead to wellbeing. It risks normalising regimes of datafied selves where technological corporations that rely on commercial gain dwell outside and within our bodies, including tethered to our nerves. Indeed, most apps and wearables that claim to address mental health emphasise the clinical validity of their product, demonstrating the increased proximity between the neurosciences, commercialisation, and technological corporations. What is different in the case of vagus nerve wearables is their reach into the body as a whole, thus both diffusing mental health in and harnessing data from the total body. Critical mental health research ought to explore how this amassing of data beyond the mind and the brain is not only impacting questions of privacy and surveillance but profoundly transforming our ontologies and epistemologies of mental health and ill health.

My discussion of vagus nerve wearables and their locating of mental health 'everywhere' in the body aims to not only broaden our understanding of mental health, but also it stresses that technological corporations are already generating this approach to mental health through their technologies. As noted above, many exercises that do not need to involve technology, such as deep breathing and yoga, are believed to help soothe

[5] While one could argue that this would exclude clinical diagnoses that are generally understood as more complex or acute, apps have been created to address aspects of early psychosis (PRIME) and bipolar disorder (eMoods Bipolar Mood Tracker).

the vagus nerve and promote balance as an indicator of wellbeing. Yet, accompanied by vagus nerve wearables, attunement involves enhancing and technologically augmenting and refining this process. At the core of these technologies lies a different onto-epistemology of mental health and ill health, one that positions mental health as matter for the total body in its social and cultural milieus.

References

Apollo Neuro. (2023, October 16). Understanding the vagus nerve: Anatomy and basics. *Apollo Neuro*. https://apolloneuro.com/blogs/news/vagus-nerve-101

Apollo Neuro. (2024). Apollo. *Apollo*. https://apolloneuro.com/

Apollo team. (2022, October 3). 4 vagus nerve exercises to transform how you handle stress. *Apollo*. https://apolloneuro.com/blogs/news/4-vagus-nerve-exercises-to-transform-how-you-handle-stress#:~:text=By%20stimulating%20the%20parasympathetic%20nervous,state%20of%20rejuvenation%20and%20recovery

Barad, K. M. (2007). *Meeting the universe halfway: Quantum physics and the entanglement of matter and meaning*. Duke University Press.

Beard, G. M. (1881). *American nervousness, its causes and consequences*. Putnam's Sons.

Borovikova, L. V., Ivanova, S., Zhang, M., Yang, H., Botchkina, G. I., Watkins, L. R., Wang, H., Abumrad, N., Eaton, J. W., & Tracey, K. J. (2000). Vagus nerve stimulation attenuates the systemic inflammatory response to endotoxin. *Nature, 405*(6785), 458–462. https://doi.org/10.1038/35013070

Charcot, J.-M. (1875). *Leçons sur les maladies du système nerveux : Faites à la Salpêtrière*. Tome 1. A. Delahaye.

Cheng, S., Chang, C., Chang, W., Wang, H., Liang, C., Kishimoto, T., Chang, J. P., Kuo, J. S., & Su, K. (2023). The now and future of ChatGPT and GPT in psychiatry. *Psychiatry and Clinical Neurosciences, 77*(11), 592–596. https://doi.org/10.1111/pcn.13588

Conrad, P. (1994). Wellness as virtue: Morality and the pursuit of health. *Culture, Medicine, and Psychiatry, 18*, 385–401.

Corning, J. L. (1882). Prolonged instrumental compression of the primitive carotid artery as a therapeutic agent. *Medical Record, 21*, 173–174.

Cosgrove, L., Naudet, F., Högberg, G., Shaughnessy, A. F., & Cristea, I. A. (2021). Reconceptualising treatment-resistant depression as difficult-to-treat depression. *The Lancet Psychiatry, 8*(1), 11–13. https://doi.org/10.1016/S2215-0366(20)30416-8

Crosby, L., & Bonnington, O. (2020). Experiences and implications of smartphone apps for depression and anxiety. *Sociology of Health & Illness, 42*(4), 925–942. https://doi.org/10.1111/1467-9566.13076

Davis, N. (2023, March 22). Gut feelings: Why drugs that nurture your microbes could be the future of mental health. *The Guardian*. https://www.theguardian.com/lifeandstyle/2023/mar/21/gut-feelings-why-drugs-that-nurture-your-microbes-could-be-the-future-of-mental-health

Dryden, J. (2023). The gut microbiome and the imperative of normalcy. *IJFAB: International Journal of Feminist Approaches to Bioethics, 16*(1), 131–162. https://doi.org/10.3138/ijfab-2022-0005

Dunn, J., Runge, R., & Snyder, M. (2018). Wearables and the medical revolution. *Personalized Medicine, 15*(5), 429–448. https://doi.org/10.2217/pme-2018-0044

Esmonde, K., & Jette, S. (2020). Assembling the 'Fitbit subject': A Foucauldian-sociomaterialist examination of social class, gender and self-surveillance on Fitbit community message boards. *Health: An Interdisciplinary Journal for the Social Study of Health, Illness and Medicine, 24*(3), 299–314. https://doi.org/10.1177/1363459318800166

Faller, A., & Schuenke, M. (2004). *The human body: An introduction to structure and function*. Thieme Medical Publishers, Incorporated.

Flore, J. (2021). Ingestible sensors, data, and pharmaceuticals: Subjectivity in the era of digital mental health. *New Media & Society, 23*(7), 2034–2051. https://doi.org/10.1177/1461444820931024

Flore, J. (2023). *The artefacts of digital mental health*. Palgrave Macmillan.

Flore, J. (2024). (Dis)assembling mental health through apps: The sociomaterialities of young adults' experiences. *Media International Australia, 192*(1), 3–20. https://doi.org/10.1177/1329878X221114486

Food and Drug Administration. (2005). Premarket Approval (PMA). Food and Drug Administration. https://www.accessdata.fda.gov/scripts/cdrh/cfdocs/cfpma/pma.cfm?id=p970003s050#:~:text=Approval%20Order-,Approval%20Order,-Summary

Frances, A. (2014). *Saving Normal: An insider's revolt against out-of-control psychiatric diagnosis, DSM-5, big pharma, and the medicalization of ordinary life*. William Morrow / HarperCollins Publishers.

Frankel, M., & Warren, M. (2023, January 23). How gut bacteria are controlling your brain. *BBC Future*. https://www.bbc.com/future/article/20230120-how-gut-bacteria-are-controlling-your-brain

Fullagar, S., O'Brien, W., & Pavlidis, A. (2019). *Feminism and a vital politics of depression and recovery*. Springer International Publishing.

Fullagar, S., Rich, E., Francombe-Webb, J., & Maturo, A. (2017). Digital ecologies of youth mental health: Apps, therapeutic publics and pedagogy as affective arrangements. *Social Sciences, 6*(4), 135. https://doi.org/10.3390/socsci6040135

Geddes, L. (2023, August 23). The key to depression, obesity, alcoholism—And more? Why the vagus nerve is so exciting to scientists. *The Guardian*. https://www.theguardian.com/society/2023/aug/23/the-key-to-depression-obesity-alcoholism-and-more-why-the-vagus-nerve-is-so-exciting-to-scientists

Gorman, R., & LeFrançois, B. A. (2017). Mad studies. In B. M. Z. Cohen (Ed.), *Routledge international handbook of critical mental health* (pp. 107–114). Routledge.

Hendry, N. A. (2020). New ways of seeing: Tumblr, young people, and mental illness. In A. McCracken, A. Cho, L. Stein, & I. Neill Hoch (Eds.), *A Tumblr book: Platform and cultures* (pp. 315–326). University of Michigan Press.

Hendry, N. A. (2023). 'Hey lovely! Don't miss this opportunity!' Digital temporalities of wellness culture, email marketing, and the promise of abundance. *Journal of Sociology, 59*(3), 664–681. https://doi.org/10.1177/14407833221101397

Heneghan, C. J., Statan, A., & Charlesworth, J. A. (2021, December 7). *Determining mental health and cognitive state through physiological and other non-invasively obtained data.* USA Patent US 11,191,466 B1.

Larkin, T. (2023, August 26). The vagus nerves help us rest, digest and restore. Can you really reset them to feel better? *ABC News*. https://www.abc.net.au/news/2023-08-26/vagus-nerves-rest-digest-nervous-system-regulation/102771178

LivaNova. (2021). SYMMETRY. Love life, live life. *LivaNova*. https://www.livanova.com/depression/getmedia/af6f9d71-5abe-41e8-bb14-cb0fb9c1293e/im-7500052-dep-symmetry-patient-brochure.pdf

LivaNova. (2023). A different approach to difficult to treat depression. https://www.livanova.com/depression/en-us

Lucas, G. (2018). Gut thinking: The gut microbiome and mental health beyond the head. *Microbial Ecology in Health and Disease, 29*(2), 1548250. https://doi.org/10.1080/16512235.2018.1548250

Lupton, D. (2019). Toward a more-than-human analysis of digital health: Inspirations from feminist new materialism. *Qualitative Health Research, 29*(14), 1998–2009. https://doi.org/10.1177/1049732319833368

May, C., Gask, L., Atkinson, T., Ellis, N., Mair, F., & Esmail, A. (2001). Resisting and promoting new technologies in clinical practice: The case of telepsychiatry. *Social Science & Medicine, 52*(12), 1889–1901. https://doi.org/10.1016/S0277-9536(00)00305-1

McAllister-Williams, R. H., Arango, C., Blier, P., Demyttenaere, K., Falkai, P., Gorwood, P., Hopwood, M., Javed, A., Kasper, S., Malhi, G. S., Soares, J. C., Vieta, E., Young, A. H., Papadopoulos, A., & Rush, A. J. (2021). Reconceptualising treatment-resistant depression as difficult-to-treat depression. *The Lancet Psychiatry, 8*(1), 14–15. https://doi.org/10.1016/S2215-0366(20)30516-2

Moncrieff, J. (2008). *The myth of the chemical cure: A critique of psychiatric drug treatment*. Palgrave Macmillan.

Moncrieff, J. (2013). *The bitterest pills: The troubling story of antipsychotic drugs*. Palgrave Macmillan.

Mörkl, S., Butler, M. I., & Wagner-Skacel, J. (2023). Gut-brain-crosstalk- the vagus nerve and the microbiota-gut-brain axis in depression. A narrative review. *Journal of Affective Disorders Reports, 13*, 100607. https://doi.org/10.1016/j.jadr.2023.100607

Moroșanu, V., Roman-Filip, I., Petruțiu, S., & Bălașa, R. (2023). Vagus Nerve Stimulation (VNS) Therapy System in pharmacoresistant epilepsy: A literature review. *Acta Marisiensis—Seria Medica, 69*(3), 157–162. https://doi.org/10.2478/amma-2023-0027

Parasym. (2024a). Mission: Restore our collective health. *Nurosym*. https://nurosym.com/pages/about

Parasym. (2024b). Nurosym. *Nurosym*. https://nurosym.com/products/nurosym

Pavlidis, A. (2023). "Troubled" derby subjectivities: Wellbeing and feminist new materialist movements in sport. *Feminism & Psychology, 33*(1), 65–82. https://doi.org/10.1177/09593535221104878

Pedersen, I. (2020). Will the body become a platform? Body networks, datafied bodies, and AI futures. In I. Pedersen & A. Iliadis (Eds.), *Embodied computing: Wearables, implantables, embeddables, ingestibles* (pp. 21–48). The MIT Press.

Pedersen, I., & Iliadis, A. (2020). Introduction: Embodied computing. In I. Pedersen & A. Iliadis (Eds.), *Embodied computing: Wearables, implantables, embeddables, ingestibles* (pp. ix–xxxix). The MIT Press.

Perez, E., & Abdel-Ghaffar, S. (2023, June 2). The Keyword. How We Trained Fitbit's Body Response Feature to Detect Stress. https://blog.google/products/fitbit/how-we-trained-fitbits-body-response-feature-to-detect-stress/

Sensate. (2024a). Product. *Sensate.* https://www.getsensate.com/products/sensate-2?variant=40503515250777

Sensate. (2024b). Science of Sensate. *Sensate.* https://www.getsensate.com/pages/science-of-sensate

Shmerling, R. H. (2021, September 22). Harvard Health. Harvard Health Ad Watch: Can a Wearable Device Reduce Stress? https://www.health.harvard.edu/blog/harvard-health-ad-watch-can-a-wearable-device-reduce-stress-202109222601

Trabsky, M., & Flore, J. (2024). Prescription medicine and economies of death. In P. Crofts (Ed.), *Evil corporations: Law, culpability and regulation*. Routledge.

Trent, J. C. (1946). Spinal anaesthesia, 1846–1946. *Journal of the History of Medicine and Allied Sciences, 1*(4), 505–514.

Tseris, E. J., Bright Hart, E., & Franks, S. (2022). "My voice was discounted the whole way through": A gendered analysis of women's experiences of involuntary mental health treatment. *Affilia, 37*(4), 645–663. https://doi.org/10.1177/08861099221108714

Wilson, E. A. (2015). *Gut feminism*. Duke University Press.

Zeavin, H. (2021). *The distance cure: A history of teletherapy*. The MIT Press.

9

Is It *Real*? Making and Unmaking Pornography Addiction

Kris Taylor

Introduction

> One question that is being asked is what actually is pornography addiction? In particular, many people may wonder if this really is an addiction, how common is it and how does it relate to 'normal' pornography use. (Adamson, 2016, n.p.)

In conversations about pornography research, I am often asked if pornography addiction is "real". Despite its superficial simplicity, to me, this question of whether pornography addiction is "real" illustrates a fundamental unease about the status of the concept. To ask whether this or that entity is "real" necessarily requires a subsequent specification: "a real *what*?". Following Austin (1962; also Hacking, 1995), the claim to something being "real" only follows from the possibility that this same

K. Taylor (✉)
The University of Auckland, Auckland, New Zealand
e-mail: ktay408@aucklanduni.ac.nz

N. A. Hendry, E. Karageorgos (eds.), *Critical Mental Health in Australia and Aotearoa New Zealand*, The Politics of Mental Health and Illness,
https://doi.org/10.1007/978-3-031-82861-4_9

something also promises to be, in some sense, *not* real. This is not to say that it does not exist, but that the "something" under description is not an authentic version of what is being described: to ask if somebody is wearing a coat made of "real fur" is only comprehensible if we understand that there is the possibility that what we perceive as fur may *not* really be from an animal. This is not the same as saying that the coat or the fur does not exist but is to enquire about the specific nature of the coat's construction. Accordingly, when someone asks me if pornography addiction is "real", I understand that the questioner is not asking whether pornography addiction exists or not but is asking about the validity of the concept. My reply is that pornography addiction is indeed real, but not in the sense that I am routinely asked.

As I explore hereafter, pornography addiction is often described in ways that illustrate the curious case of the concept as a product of both popular appeal and a crisis of legitimacy. By way of introduction, consider the following extract from an article in *The New Zealand Herald*, titled *Can you really be addicted to pornography?* (Adamson, 2016). In this article, Associate Professor Simon Adamson discusses two-time Olympic medallist Nick Willis's public confession of a self-diagnosed pornography addiction. However, he does so in a way that elegantly illustrates a tension between pornography addiction's illegitimacy on the one hand, and its popularity on the other:

> Firstly, the official manual of the American Psychiatric Association – which is also used in New Zealand – does not recognise pornography addiction. Gambling is the only behavioural addiction to receive official sanction for [sic] that organisation. In my practice as a clinical psychologist specialising in addiction, I see people for a range of substance problems. The most common non-substance behaviour that people seek my assistance with is pornography addiction, with many clients frequently using this term when first making contact. When I assess a person, it is important to keep an open mind about what is occurring for them. However I have consistently found that these men (and they have all been men) describe problems with compulsive pornography use that closely mirrors most of the symptoms of substance addiction. Furthermore, interventions that have been developed to work with substance misuse have seemed generally well suited to assisting these men recover from their pornography problem.

Adamson's account of pornography addiction's validity here suggests an ambiguous ontology. First, he suggests that the diagnosis is not legitimate, as the American Psychiatric Association does not recognise it. He then describes the popularity of the pornography addiction concept among those who seek him out for treatment before going on to describe the behaviours of his clients as presenting in similar ways to substance addiction and recommending treatments as such. Thus, on the one hand, when taken at face value, it seems obvious that pornography addiction *is* real: Adamson describes pornography addiction as "real" enough for some (male) pornography viewers to describe themselves in such a way and real enough for the application of treatments based on substance abuse. On the other hand, however, this claim to pornography addiction's 'realness' is undercut by the fact that pornography addiction remains unrecognised "officially" as a discrete diagnosis or disorder. Pornography addiction remains absent as a diagnosis in both the fifth edition of the *Diagnostic and Statistical Manual* (DSM-5) of the American Psychiatric Association (2013; see Weir, 2014) and the eleventh revision of the *International Classification of Diseases* (ICD-11; see Kraus et al., 2018; Montgomery-Graham et al., 2015). In this sense, if the question is whether pornography addiction is a "real" psychiatric disorder, calling specifically on the authority of "official" recognition and evidence, then in this article, Adamson is also describing pornography addiction as somehow "unreal"—or at least only provisionally real—by virtue of its failure to reach designation in the DSM-5.

The example of Adamson's (2016) article establishes the central tension explored in the following commentary. While there is certainly a freely circulating lay understanding of pornography addiction used by some people to describe various forms of problematic pornography viewing, this is not to say that the diagnosis is 'valid' (although the question of validity raises questions of where, when, and to whom a diagnosis is valid or not). Thus, there is a tension between the status of pornography addiction as a diagnosis in some clinical sense and pornography addiction as a broader social construct. Indeed, in reference to the broad context of pornography addiction's popularity, it is worth noting that although Adamson is writing from his context as a clinical psychologist in Aotearoa New Zealand, the conceptual framing and the accounts of his patients are indistinguishable from popular discussion of pornography addiction, regardless of date of publication, or whether from the USA (see Borrensen, 2019; Skinner, 2014) or UK (see Blunden, 2018; Morrison, 2023; Wendling, 2016).

In this chapter, I expand upon the tenuous ontological status of pornography addiction just described, hovering as it does between contested evidential base and popular social explanation. The following commentary is most indebted to Hacking's (1986, 1995) work, which reconceptualises how categories arise within antecedent fields of knowledge and how new categorisations of people invariably shape the experiences of those within that category. In what follows, my interest is in identifying ways that flows of knowledge between categoriser and categorised co-create new understandings of behaviour (see also Brossard et al., 2023; Taylor, 2019, 2020). As such, I will begin by discussing the search for the 'official' (to borrow Adamson's phrasing) status of pornography addiction using examples from scholarly literature on pornography, describing some of the key research used as evidence for the 'realness' of the diagnosis. In doing so, I will highlight how an 'official' pornography addiction diagnosis can only be vaguely made out, buried as it is under larger, more established fields of research on excessive behaviours and pathological sexuality. Thus, instead of looking at the validity of this literature *per se* (or offering a comprehensive review), I aim to interrogate how empirical attempts to 'make' pornography addiction an official taxonomy work in tandem with lay discourses to make the concept of "addictive" pornography viewing popular.

However, as described above, new categories are not simply 'introduced' but are developed and reshaped through the input of those under categorisation (Hacking, 1986). Accordingly, understanding empirical approaches to pornography addiction is only part of the story. Having briefly highlighted some key empirical research that helps to construct a 'real' pornography addiction (mainly from a psychological research perspective), I will turn to the "social life" of the concept. Here, the discussion shifts from the search for a 'real' pornography addiction in an empirical sense towards the rhetorical appeal of the idea in popular culture more broadly. That is, while the official pornography addiction concept is being conceptualised, operationalised, measured, and debated inside of academic jurisdictions, it is also conceptualised, operationalised, measured, and debated in ways untethered from any 'official' authority. While such an approach resonates with previous work on sex addiction by Irvine (1995), Reay et al. (2013) and Brossard et al. (2023), the specifics of how diagnoses become popularised outside of professional jurisdictions require further consideration (Brossard & Chandler, 2022). This consideration is especially the case for pornography addiction because, as I will explore, comparisons between the "official" status of pornography addiction and the popularity of the diagnosis in popular media indicate that pornography addiction's "realness" arises irrespective of the lack of consensus about pornography's proposed effects on its audience. Thus, we must look more closely at the pornography addiction concept as co-constructed somewhere between expert and lay fields of knowledge. It is this space between an official and unofficial diagnosis, between the real and unreal, and between expert and lay fields of knowledge that I propose pornography addiction is being *made* real.

A Real Disorder? Scholarly Research on Pornography Addiction

The content of this section is not intended to be a 'literature review' of pornography addiction in any formal sense because, not only would such a task require considerably more historical exploration and theorising than I have room for here (see Taylor, 2019), but this work has also already been done (better) elsewhere (Burke & MillerMacPhee, 2021; Kohut et al., 2019; Montgomery-Graham et al., 2015). Instead, here I want to highlight some of the conceptual issues that arise when searching for the empirical basis of pornography addiction. Indeed, here I would ideally focus my enquiry into the 'official' designation of pornography addiction on a body of research on "pornography addiction" specifically. That is, my focus would ideally be on research working to investigate an 'addiction' (as opposed to dysregulated, compulsive, impulsive, or otherwise problematic viewing) to 'pornography' (as opposed to an addiction to the internet, cybersex, masturbation, sex, or other sexual stimuli).

However, as we will see, such a demarcation is difficult to make out or maintain. As a case in point, it is worth noting that in attempting to partition off a specific branch of pornography addiction research, we are already on shaky taxonomic ground because conceptualising 'pornography addiction research' as a sovereign scholarly territory is complicated by the presence of other, adjacent research traditions. It is a stark observation that the field of sex research already has a rich stockpile of viable taxonomies to describe problematic pornography viewing: "hypersexuality, sex addiction, sexual compulsivity, sexual impulsivity, dysregulated sexual behaviour, [and] out-of-control sexual behaviour" all enjoy some legitimacy in contemporary psychological literature (Walton et al., 2017, p. 2231; see Ley & Grubbs, 2017 for commentary), and each of which could feasibly describe problematic pornography viewing within their purview. This crowded field of taxonomies presents a serious challenge to recognising the authority of a pornography addiction diagnosis as legitimate, in the sense that pornography addiction is a minor player in the fields of pornography and sex research.

Whether the literature on these other conceptualisations of problematic sexuality should be described as evidence for or against, the existence of a standalone pornography addiction taxonomy rapidly becomes an almost philosophical exercise in semantics. For example, although some might argue that research on 'compulsive sexual behaviour' or 'internet addiction' can act as evidence that pornography is addictive (e.g., that internet addiction is spurred by watching pornography or that a symptom of compulsive sexual behaviour is pornography viewing), such an argument intrinsically acknowledges that the label "pornography addiction" is not widely employed by researchers. Where "compulsive cybersex" could be interpreted as an analogue to pornography addiction, compulsive cybersex is a field unto itself, also struggling to eke out its own unique interpretation of problematic pornography viewing (see Cooper et al., 2000; Wéry & Billieux, 2017). Thus, the term "pornography addiction literature" is used advisedly here because besides research under other tangentially related 'addictions' (i.e., cybersex addiction, sex addiction, internet addiction, internet sex addiction, etc.), empirical research on "pornography addiction" *specifically* (i.e., attempts to measure an addiction to pornography) is relatively small. Indeed, a review of the extant literature has suggested that as of 2021, there were roughly 127 scientific articles that mention pornography addiction somewhere in the text, with only 103 included in the final review (Burke & MillerMacPhee, 2021). Of Burke and MillerMacPhee's corpus, 21 articles were review articles, and 11 were brain imaging studies, with most pornography research being questionnaire-based. Considering the centrality of descriptions of the action of neurotransmitters in popular conceptions of pornography addiction (i.e., dopamine), it is interesting to note that this corpus contains only a handful of brain imaging studies (e.g., Brand et al., 2016; Gola et al., 2017; Kühn & Gallinat, 2014; Prause et al., 2015; Voon et al., 2014). Further, considering the longevity of the pornography addiction concept (going back to at least the mid-1980s; see Taylor, 2019), it might be surprising to note that the bulk of research on pornography addiction has only been carried out in the last decade (Burke & MillerMacPhee, 2021).

Crucially for understanding the shape of an 'official' pornography addiction diagnosis, the conceptualisation and measurement of pornography addiction are at times methodologically inconsistent and theoretically incoherent (Binnie & Reavey, 2019; Wéry et al., 2019). For example, where some pornography addiction literature focuses on the action of dopamine and the reward pathway specifically (Hilton, 2013; Kühn & Gallinat, 2014), others bypass mentioning the action of dopamine almost entirely (Brand et al., 2016; Gola et al., 2016). Inversely, where some rely on concepts of withdrawal and tolerance (Ford et al., 2012), others ignore these criteria (Hilton, 2013). Where some research suggests a similarity in brain activity between people with self-described "compulsive sexual behaviours" being shown pornography to dependent drug users being shown drug stimuli (Voon et al., 2014), other research suggests the opposite, with problematic viewers showing decreased activation (Prause et al., 2015). Pertinently, in the case of Voon et al.'s (2014) research, which focused on men with "compulsive sexual behaviours" (raising questions as to whether this should even be included in a review of pornography addiction literature as discussed above), the authors have publicly cautioned against such an interpretation (Borrensen, 2019). Yet, despite such cautioning, this study is routinely cited by academics, researchers, and lay people as evidence for the existence of a pornography 'addiction' (e.g., Brand et al., 2016; Gola et al., 2017; Love et al., 2015; Sniewski et al., 2018). For example, as a story in *The Independent* (UK) describes:

> People who are addicted to pornography show similar brain activity to alcoholics or drug addicts, a study has revealed. MRI scans of test subjects who admitted to compulsive pornography use showed that the reward centres of the brain reacted to seeing explicit material in the same way as an alcoholic's might on seeing a drinks advert. (Withnall, 2013, n.p.)

Arguably, this discrepancy between the stated target of Voon et al.'s (2014, p. 9) influential research—which mentions the word "pornography" once—and its subsequent interpretation as proof of pornography addiction is a discrete and measurable brain state that exemplifies the rhetorical appeal of pornography addiction as superseding available evidence.

Accordingly, a central problem for pornography addiction research is the mismatch between the casual lay usage of the addiction concept (e.g., "I'm addicted to Sinatra" or "He's a golf addict"; see Mihordin, 2012, p. 489; Taylor, 2020) and the operationalisation and measurement of an apparently discrete disease state built on that same taxonomy. For example, the liberal application of 'addiction' in popular culture (as explored further below) has led to ongoing academic debate about whether 'addiction' is a useful diagnostic category. As Schmitz (2005, p. 153) suggests: "the term 'addiction' has been overused and misused in society to the point that its scientific value is limited", and the expansion of the addiction concept in research has been criticised as risking "a severe over-pathologisation of everyday behaviours" (Billieux et al., 2015, p. 120; see also Conrad & Schneider, 1992; Keane, 2004: Mihordin, 2012; Reinarman, 2005; Room, 2014; Walters & Gilbert, 2000). Crucially, in the case of the pornography addiction concept specifically, the usefulness of an 'addiction' framing has been described in a therapeutic context as "a simplification of a complex individual's psychological functioning with limited clinical relevance" (Wéry et al., 2019, p. 124). Indeed, at a broader level, there are a series of important questions to ask about how to accurately conceptualise the measurement of addictions to the internet, sex, cybersex, and pornography: Is it masturbation, pornography, or the internet that activates such addiction (see Prause, 2019)? Is it the content, the frequency, or the duration of pornography viewing that is indicative of addiction (see Gola et al., 2016)? Does collecting increasingly large amounts of pornography (Hall, 2013) or having multiple browser tabs open at one time (Blunden, 2018) indicate an addiction-like tolerance? How do other 'non-pathological' activities, including watching sports, viewing images of sweet foods, and recognising brands of chocolate (which all stimulate neural responses similar to those interpreted as evidence for pornography's addictiveness) compare (see Ley et al., 2014)? Following Moser (2011, p. 228), the distinctly rhetorical appeal of a "pornography addiction" is further complicated by a consideration of how such a diagnosis raises interesting dilemmas of nosology: "Individuals who obsessively or compulsively wash their hands may have an obsessive-compulsive disorder (OCD), but they do not have a hand-washing disorder". In this sense, is what some describe as a "pornography addiction" a discrete and diagnostically useful taxonomy, or is it a symptom of some broader diagnostic category related to impulse control?

All of which is to say, the 'realness' of pornography addiction remains a contentious issue among researchers (see Clarkson & Kopaczewski, 2013; Giugliano, 2013; Ley et al., 2014; Spišák, 2016; Wéry & Billieux, 2017). As this brief review of the relatively small body of literature on "pornography addiction" suggests, despite the seeming popularity of the concept in popular culture, the 'official' status of pornography addiction as a physiological state, a clinical diagnosis, or some other unique construct of the psychological sciences remains dubious. In this sense, while an addiction framing may have an authoritative rhetorical appeal, this has not as yet been convincingly translated into the field of pornography or sex research. Accordingly, we must ask why the seemingly awkward and conceptually fuzzy framing of 'pornography addiction' persists. One suggestion is that perhaps the distinctly fluid contemporary morphology of the addiction concept (Keane, 2004) holds a sustained appeal because such a framing promises to unite otherwise disparate understandings of problematic pornography viewing under a single taxonomic rubric. In other words, dysregulated, compulsive, impulsive, or otherwise problematic pornography viewing could be brought under an addiction taxonomy and thus simplify the field of research on problematic pornography viewing. In this sense, 'addiction' would work as a shorthand for describing problems of excess and losing self-control without needing to attend to the social conditions or circumstances in which this apparent loss of control occurs (see Grubbs & Perry, 2019; McKee, 2007a; Taylor, 2023). It is this rhetorical appeal to which we turn next, with an eye towards the ways that the presence of pornography addiction within academic literature is itself a reflection of the social life of the pornography addiction concept rather than a discrete and measurable disease state.

A Real Problem? The Social Life of Pornography Addiction

> In *The Porn Circuit*, you learned how neurochemicals like dopamine and testosterone are activated when a person views porn and how they work together to build an addiction. (Covenanteyes.com, 2019)

The above is a quote taken from a series of emails that I received after downloading a resource from the pornography addiction self-help website Covenant Eyes (USA), which boasts some 1.5 million app users (www.covenanteyes.com). Considering the discussion above about pornography addiction research, what I find interesting about this quote is how authoritative its claims are. This quote illustrates the way that the uncertainties of conceptualising pornography as addictive in research are readily translated into compelling pronouncements about how pornography *is* addictive outside of a research context. This pronouncement is illustrative of the explanatory rhetorical appeal of the pornography addiction concept: the pornography addiction diagnosis exists not because of a scientific consensus about the criteria for the addiction but because the concept combines a culturally meaningful explanation with a thin veneer of medical authority. Despite pornography addiction not being officially recognised by many researchers or mental health diagnostic bodies like the DSM and ICD, it persists as a seemingly common-sense way to describe some pornography viewing/viewers. In this sense, even though reviewing research on pornography addiction indicates a lack of conceptual clarity, we can see that, in some sense, the fuzziness of the concept in this research does not really matter. Thus, despite acknowledging the tenuous clinical status of pornography addiction above, the original question of pornography addiction's 'realness' must be transformed into a question of how we otherwise 'make it up' (Hacking, 1986). That is, if pornography addiction is not empirically derived *per se*, then how is pornography addiction *socially* derived?

While pornography addiction may not be validated as 'real' in an empirical sense, pornography addiction has certainly become a popular way to describe problematic pornography-adjacent behaviours in popular culture. Today pornography addiction circulates between social jurisdictions, a curiously versatile diagnosis shared by celebrities (e.g., Russell Brand, Terry Crews, David Duchovny, Jada Pinkett Smith, Kanye West) and the otherwise ordinary pornography viewers described by Adamson (2016), along with rapists (Goldman, 2013) and paedophiles (McKenna, 2019). That pornography addiction can be used to describe the behaviours of these presumably disparate groups tells us something about the peculiar sociocultural utility of the concept, describing behaviours ranging from the seemingly frivolous to the criminal. In the first instance, consider the release of HBO's *Mrs Fletcher*, a television series about the sexual awakening of a "porn-addict mom" (D'Addario, 2019) upon her discovery of so-called MILF ("mother I'd like to fuck") pornography. Or else the confession of actor Jada Pinkett Smith of her "little porn addiction", a term she uses colloquially to describe her previous "unhealthy" relationship with pornography (Borrensen, 2019, n.p.). Yet, the use of the pornography addiction diagnosis in these ways is made incongruous when set beside the story of a man "provisionally diagnosed with a pornography addiction and paedophilia disorder" standing trial for the kidnapping and subsequent sexual abuse of a seven-year-old (McKenna, 2019, n.p.), or the account of a 17-year-old whose "pornography addiction" ostensibly led him to repeatedly rape his then 14-year-old girlfriend (New Zealand Herald, 2019). While the application of a pornography addiction diagnosis could arguably be intuited as manifesting differentially between Olympian Nick Willis and serial killer-cum-self-diagnosed pornography addict Ted Bundy (who described his exposure to pornography as the basis for his crimes prior to his execution), it is precisely this contextual differentiation that interests me hereafter: the use of the same word to describe a wide array of problematic behaviours related to pornography viewing.

Where empirical efforts to quantify pornography addiction have stumbled over how to conceptualise, operationalise, and measure pornography as addictive as described above (i.e., frequency, duration, content), the rhetorical appeal of popular pornography addiction discourse persists despite this evidential vacuum. Take, for example, the routine tendency of news media reporting to both acknowledge pornography addiction's uncertain official diagnostic status while also describing this lack of designation as ancillary to the existence of a 'real' pornography addiction (self)diagnosis: "Whatever scientists might say, Tom feels he's addicted to porn" suggest the BBC (Wendling, 2016, n.p. see also Adamson, 2016; Blunden, 2018, Fidgen, 2013; Kelly, 2017; MacDonald, 2016; Skinner, 2014; Weir, 2014). Indeed, there is something telling about the ways that some of these articles are titled, often framed as a question, that betrays an acknowledgement of a tension between a lack of evidence on the one hand and the popular circulation of the concept on the other: "Is porn addiction real?" (Borrensen, 2019); "Is pornography (sexual) addiction real?" (Skinner, 2014); "Is 'porn addiction' a real thing?" (Wendling, 2016); "Is pornography addictive?" (Weir, 2014). A recent article in *The Daily Mail* (Morrison, 2023) titled "Is porn addiction real? How to tell if you have it and what to do about it" illustrates this tendency well by both querying and affirming the veracity of pornography addiction at the same time, in the title and article itself.

However, rather than focusing on how media reporting makes sense of the criteria for problematic pornography viewing (see Taylor & Gavey, 2020), here I want to address a rarely asked question that is hinted at throughout such reporting about pornography addiction: what does 'normal' pornography viewing actually look like? That is, fundamentally, the concept of an addiction to pornography suggests the existence of an unspecified threshold between acceptable and unacceptable pornography viewing. Accordingly, media reporting routinely acknowledges the difficulties of delineating between what is understood as 'normal' and 'problematic' pornography viewing:

> While compulsive porn use may not be a diagnosable addiction by the DSM's standards, there's no doubt that for some, it's a real issue with potentially serious and negative consequences. But how do you know when porn use has crossed the line into potentially unhealthy territory? (Borrensen, 2019, n.p.)

Borrensen's (2019) query—"How do you know when porn use has crossed the line into potentially unhealthy territory?"—provokes a necessary question as to *where* the line between addictive and non-addictive viewing might lie. Such reporting suggests a fascinating tension between this lack of official diagnostic criteria and the presence of some ambiguous point of threshold at which pornography viewing ostensibly *becomes* addictive. Thus, instead of beginning by asking a question like "How much porn is too much?" I argue that we must first contextualise this question against the presence of an undefined 'correct' or 'healthy' amount of pornography being viewed in the first place. Indeed, a somewhat ironic cliché employed in news articles working to describe the dangers of pornography addiction is that these dangers are frequently set alongside disclaimers about the normality of such viewing:

> It is probably safe to say that the majority of Americans (myself included) have viewed some pornography when were were [sic] relatively young, however minimal, without having it dramatically alter our lives. That being said, the downside is that in our current environment, too many people (men in particular) are seeing their lives turned upside down as pornography has increasingly gained considerable control of their wellbeing — sexually, psychologically, socially, financially and in some cases, physically. (Watson, 2014, n.p.)

Such accounts illustrate a fascinating attempt to balance the widespread, 'normal' viewing of pornography with the ostensibly catastrophic effects of pornography on its viewers. Similarly, in the epigram that begins this chapter, Adamson (2016) asks how pornography addiction compares to "normal" pornography viewing, creating both a baseline from which a viewer might deviate and introducing a spectre of what that deviation might look like. Here, pornography addiction acts as an ill-defined dividing line between acceptable and unacceptable pornography viewing. And as Adamson (2016) reassures the reader in the second-to-last sentence of his article:

> My clinical experience tells me that compulsive pornography use can helpfully be understood and treated, as an addiction. This does not mean all pornography use is problematic.

It strikes me as peculiar in these cases that while pornography viewing is described as hazardous, it is also routinely—and often concurrently—described as normal, if not a normative and healthy part of sexuality, especially for men and boys (see Boyle, 2010; Flood, 2008; García-Favaro, 2015; McKee, 2007b; also Office of Film and Literature Classification, 2018). As Karen Boyle (2010) has argued, pornography viewing has largely become normalised in popular culture. Whether in the self-deprecating humour of so-called lad-mags, the visibility of references to pornography brands in films and television, or the centrality of pornography viewing in representations of teenage life in popular culture, pornography viewing is mainstream: "Porn use is an in-joke, a homosocial experience, a 'natural' expression of youthful sexuality, even a mark of distinction and source of cultural capital" (Boyle, 2010, p. 144). Thus, to understand the popular appeal of the pornography addiction concept, we need to realise that pornography addiction is necessarily set against a cultural backdrop of an acceptable level of pornography viewing that is as poorly defined as when pornography viewing goes wrong.

In turn, this normalisation of pornography viewing can make some critiques of pornography (e.g., moral, ethical, or political) more difficult to articulate than others (e.g., medical) without the risk of backlash. As Nikunen (2007) describes in an analysis of women's discussions of pornography online, criticisms of pornography can be made difficult by countervailing discourses of pornography as fun, educational, or as a sex aid, thereby minimising experiences of disturbance or disgust (see Paasonen, 2007; Parvez, 2006). According to García-Favaro (2015, p. 368), pornography viewing might not only be seen in these ways but as a biological imperative for men, explained as a "fact of life" for women to compete with. Accordingly, we can start to make out the perimeters of a discursive vacuum that surrounds pornography viewing generally: while some viewing is acceptable, and some is not, articulating a sense of unease with the presence of pornography in one's life is made difficult by contrary messages that pornography viewing is normal, and in the case of young men normative.

I argue that despite its shaky empirical foundations, the popular appeal of the pornography addiction concept is that it offers an ostensibly objective resistance to (if not a critique of) pornography. Here, following Zola (1972, p. 213), we can see how "the labels health and illness are remarkable 'depoliticisers' of an issue". A pornography addiction diagnosis solves the problem of needing to distinguish possible motives, desires, ambivalences and tensions experienced by 'users' to instead frame these issues as a singular, easy-to-understand, personal pathology. Moreover, the seemingly compressive explanatory lens of pornography addiction has changed the way that pornography viewers make sense of themselves (Grubbs & Perry, 2019; Taylor, 2020). The ambiguous authority of the pornography addiction concept creates a sort of omnipresent lens through which one must relate to both the personal and sociocultural consequences of pornography viewing. Thus, in the case of those who take up the diagnosis for themselves, we must be careful to understand that this process of self-diagnosis is dynamic and that upon inspection, the propagation of a pornography addiction diagnosis follows a process more grassroots than iatrogenic: Clearly, the pornography addiction concept represents a sort of self-problematising that happens outside of the boundaries of "official" designation. In turn, the confession of a pornography addiction acts as a sort of confession which cleaves the viewer from the social embeddedness of their viewing. If a viewer is distressed by their pornography viewing—in almost any conceivable way—then the pornography addiction diagnosis wields the kind of explanatory power to successfully describe and assuage such distress. In other words, today, pornography viewers are suspended between the poles of laissez-faire sexual and technological liberalism on the one hand and popular moral, ethical, and political concerns about pornography viewing on the other. The pornography addiction concept here offers the promise of a diagnostic panacea that will dissolve the inherent tensions between these two points of view, allowing both perspectives to coincide despite their intrinsic incompatibility.

Thus, by highlighting the normalisation of pornography viewing, the social utility of the pornography addiction concept is cast in a different light. With few discursive resources through which to address pornography viewing as a normalised and normative practice, 'pornography addiction's' explanatory rhetorical appeal makes it a versatile discursive tool. Pornography addiction glides over the complex questions of the problematics of pornography viewing while also acting as a conveniently ambiguous dividing line between pornography viewing that is acceptable and pornography viewing that is not, for whatever reason. Here, pornography addiction offers a 'third way' between a liberal tolerance towards pornography viewing and the possible viewer harms posed by pornography. In this meditating role, pornography addiction serves a largely ignored sociocultural function: foreclosing a thorough interrogation of, or serious engagement with, the widespread appeal of the very media that ostensibly causes 'addiction' in the first place.

Conclusion: (Un)making Pornography Addiction

Asking questions about how we conceptualise a popular diagnosis like pornography addiction has important implications because diagnoses carve up our social world in ways that both reflect and create new social conditions (Brossard & Chandler, 2022). For example, interrogating popular diagnoses helps us to think through why a verbal shorthand like "pornography addiction" is now used to explain almost any problematic behaviour related to pornography and further opens up a consideration of how making such self-diagnoses "real" bends towards a dismissal of ethics, morals, criminality, and so on. Crucially, it also provokes a consideration of the sociocultural context within which research programmes are formulated and how such research, in turn, works to 'make up' scientific objects out of culturally bound social concerns (Hacking, 1986). As I have argued, efforts to codify a 'real' pornography addiction in psychological literature are itself reflective of the operation of a rhetorical appeal, in that the addiction concept in popular culture has driven the establishment of efforts to find proof of its existence. Indeed, a frequently absent observation about pornography research is the acknowledgement that research is formulated within a historically and culturally contingent social milieu that has significant bearing upon both the focus and communication of such research. In the case of pornography addiction, such research frequently starts with the presumption of its 'effect' without adequately conceptualising how such effects might arise or why this particular line of inquiry is being followed in the first place (Ley et al., 2014). As Wartella and Reeves (1985; also Wilson, 1973) have shown of historical trends in media effects research on children, public concerns about new forms of media are a reliable predictor of the establishment of new fields of research inquiry. As they suggest, public debates about new forms of media shape research agendas to a greater degree than research has shaped public concerns or policy. Research by Burke and MillerMacPhee (2021) on pornography news coverage and academic research on pornography addiction shows precisely this pattern, with spikes in interest in media preceding spikes in pornography research. Accordingly, as Vörös (2009) argues, the search for an empirical basis for pornography addiction is itself based upon evidence derived from popular cultural discourse. First, therapists compile accounts of clients who describe themselves as 'addicted' to pornography, which in turn creates a corpus from which definitions and characteristics might be promulgated (e.g., Adamson describes "many clients frequently using this term [pornography addiction] when first making contact"). In turn, the presence of this corpus encourages the development of screening tests and surveys, which do the work of translating subjective client assessments about being 'addicts', in a colloquial sense, into an ostensibly value-free and empirical object of interest (see also Plummer, 1995). In turn, this object of interest is turned into a research programme with little regard for the concept's origin in lay fields of understanding. Thus, research on pornography addiction would be better understood as a struggle to realise a chimeric promise of transforming rhetorical appeal and popular discourse of concern about pornography into something that resembles a 'real' disease state.

Pornography addiction is not a physiologically measurable state waiting to be discovered by a canny neuroscientist but a discursive tool for describing almost any aspect of collective social anxiety about pornography as it currently exists. In the rhetorical sense, the evidence for pornography addiction as 'real' is superseded by the popularity of the term. While a lay diagnosis of pornography addiction relies on ambiguous criteria and is applied liberally to celebrities, rapists, and lay people alike, its utility as a catch-all term and as a shorthand that describes pornography viewing 'gone wrong' generally has made pornography addiction real in the popular sense, and almost impossible to quantify in an empirical diagnostic sense. While academic debates about the 'brains' of so-called addicts continue, in many ways, whether pornography addiction is 'real' or not is redundant. Where the accounts of clinicians (Adamson, 2016; Levine, 2010; Macdonald, 2016; Skinner, 2014) might imply that pornography addiction is spreading like an epidemic of some new disease, here pornography addiction is better viewed as the propagation of a concept with a strong rhetorical appeal. Pornography is not addictive in the sense that it represents a particular *kind* of pornography viewer whose brain has become out of control; pornography is only addictive in as much as it is popularly described as such.

References

Adamson, S. (2016, March 7). Can you really be addicted to pornography? *New Zealand Herald*. https://www.nzherald.co.nz/nz/news/article.cfm?c_id=1&objectid=11601461

American Psychiatric Association. (2013). *Diagnostic and statistical manual of mental disorders* (5th ed.). American Psychiatric Association.

Austin, J. L. (1962). *Sense and sensibilia*. Clarendon Press.

Billieux, J., Schimmenti, A., Khazaal, Y., Maurage, P., & Heeren, A. (2015). Are we overpathologising everyday life? A tenable blueprint for behavioural addiction research. *Journal of Behavioural Addictions, 4*(3), 119–123. https://doi.org/10.1556/2006.4.2015.009

Binnie, J., & Reavey, P. (2019). Problematic pornography use: Narrative review and a preliminary model. *Sexual and Relationship Therapy, 35*(2), 137–161. https://doi.org/10.1080/14681994.2019.1694142

Blunden, M. (2018, May 17). Number of Londoners seeking help for porn addiction soars. *Evening Standard*. https://www.standard.co.uk/news/health/number-of-londonersseeking-help-for-porn-addiction-soars-a3841541.html

Borrensen, K. (2019, May 22). Is porn addiction real? Here's what experts have to say. *Huff Post.* https://www.huffpost.com/entry/porn-addictionadvice_l_5ce43778e4b075a35a2e8f15

Boyle, K. (2010). Porn consumers' public faces: Mainstream media, address and representation. In *Everyday pornography* (pp. 146–158). Routledge.

Brand, M., Snagowski, J., Laier, C., & Maderwald, S. (2016). Ventral striatum activity when watching preferred pornographic pictures is correlated with symptoms of Internet pornography addiction. *NeuroImage, 129*, 224–232. https://doi.org/10.1016/j.neuroimage.2016.01.033

Brossard, B., & Chandler, A. (2022). *Explaining mental illness: Sociological perspectives.* Bristol University Press.

Brossard, B., Roy, M., Brown, J., Hemmings, B., & Larocque, E. (2023). On the social existence of mental health categories: The case of sex addiction. *Social Problems, spad024.* https://doi.org/10.1093/socpro/spad024

Burke, K., & MillerMacPhee, A. (2021). Constructing pornography addiction's harms in science, news media, and politics. *Social Forces, 99*(3), 1334–1362. https://doi.org/10.1093/socpro/spad024

Clarkson, J., & Kopaczewski, S. (2013). Pornography addiction and the medicalisation of free speech. *Journal of Communication Inquiry, 37*(2), 128–148. https://doi.org/10.1177/0196859913482330

Conrad, P., & Schneider, J. W. (1992). *Deviance and medicalisation: From badness to sickness.* Temple University Press.

Cooper, A., Delmonico, D. L., & Burg, R. (2000). Cybersex users, abusers, and compulsives: New findings and implications. *Sexual Addiction & Compulsivity: The Journal of Treatment and Prevention, 7*(1–2), 5–29. https://doi.org/10.1080/10720160008400205

D'Addario, D. (2019, September 10). TV Review: 'Mrs. Fletcher'. *Variety.* https://variety.com/2019/tv/reviews/mrs-fletcher-review-hbo-kathryn-hahn-tiff-1203330875/

Fidgen, J. (2013, June 25). Do we know whether pornography harms people? *BBC Radio 4.* https://www.bbc.com/news/magazine-22987051

Flood, M. (2008). Men, sex, and homosociality: How bonds between men shape their sexual relations with women. *Men and Masculinities, 10*(3), 339–359. https://doi.org/10.1177/1097184x06287761

Ford, J. J., Durtschi, J. A., & Franklin, D. L. (2012). Structural therapy with a couple battling pornography addiction. *The American Journal of Family Therapy, 40*(4), 336–348.

García-Favaro, L. (2015). Porn trouble: On the sexual regime and travels of postfeminist biologism. *Australian Feminist Studies, 30*(86), 366–376. https://doi.org/10.1080/08164649.2016.1150937

Giugliano, J. R. (2013). Sex addiction as a mental health diagnosis: Coming together or coming apart? *Sexologies, 22*(3), 77–80. https://doi.org/10.1016/j.sexol.2012.08.018

Gola, M., Lewczuk, K., & Skorko, M. (2016). What matters: Quantity or quality of pornography use? Psychological and behavioural factors of seeking treatment for problematic pornography use. *The Journal of Sexual Medicine, 13*(5), 815–824.

Gola, M., Wordecha, M., Sescousse, G., Lew-Starowicz, M., Kossowski, B., Wypych, M., & Marchewka, A. (2017). Can pornography be addictive? An fMRI study of men seeking treatment for problematic pornography use. *Neuropsychopharmacology, 42*(10), 2021–2031. https://doi.org/10.1038/npp.2017.78

Goldman, R. (2013, July 27). Ariel Castro accepts plea of "Not less than a thousand years". *ABC News*. https://abcnews.go.com/US/ariel-castro-accepts-dealthousandyears/story?id=19781739

Grubbs, J. B., & Perry, S. L. (2019). Moral incongruence and pornography use: A critical review and integration. *The Journal of Sex Research, 56*(1), 29–37. https://doi.org/10.1080/00224499.2018.1427204

Hacking, I. (1986). Making up people. In T. C. Heller, M. Sosna, & D. Wellbery (Eds.), *Reconstructing individualism: Autonomy, individuality, and the self in Western thought* (pp. 161–169). Stanford University Press.

Hacking, I. (1995). *Rewriting the soul: Multiple personality and the sciences of memory*. Princeton University Press.

Hall, P. (2013). A new classification model for sex addiction. *Sexual Addiction & Compulsivity, 20*(4), 279–291. https://doi.org/10.1080/10720162.2013.807484

Hilton, D. L., Jr. (2013). Pornography addiction—A supranormal stimulus considered in the context of neuroplasticity. *Socioaffective Neuroscience & Psychology, 3*(1), 20767. https://doi.org/10.3402/snp.v3i0.20767

Irvine, J. (1995). Reinventing perversion: Sex addiction and cultural anxieties. *Journal of the History of Sexuality, 5*, 429–450.

Keane, H. (2004). Disorders of desire: Addiction and problems of intimacy. *Journal of Medical Humanities, 25*(3), 189–204. https://doi.org/10.1023/b:jomh.0000036637.03254.38

Kelly, G. (2017, September 11). The scary effects of pornography: How the 21st century's acute addiction is rewiring our brains. *Telegraph*. https://www.telegraph.co.uk/men/thinking-man/scary-effects-pornography-21stcenturys-acute-addiction-rewiring/

Kohut, T., Balzarini, R. N., Fisher, W. A., Grubbs, J. B., Campbell, L., & Prause, N. (2019). Surveying pornography use: A shaky science resting on poor measurement foundations. *The Journal of Sex Research*, 1–21. https://doi.org/10.1080/00224499.2019.1695244

Kraus, S., Krueger, R., Briken, P., First, M., Stein, D., Kaplan, M., & Reed, G. M. (2018). Compulsive sexual behaviour disorder in the ICD-11. *World Psychiatry, 17*(1), 109–110.

Kühn, S., & Gallinat, J. (2014). Brain structure and functional connectivity associated with pornography consumption: The brain on porn. *JAMA Psychiatry, 71*(7), 827–834. https://doi.org/10.1001/jamapsychiatry.2014.93

Levine, S. B. (2010). What is sexual addiction? *Journal of Sex and Marital Therapy, 36*(3), 261–275. https://doi.org/10.1080/00926231003719681

Ley, D., Prause, N., & Finn, P. (2014). The emperor has no clothes: A review of the "pornography addiction" model. *Current Sexual Health Reports, 6*, 94–105. https://doi.org/10.1007/s11930-014-0016-8

Ley, D. J., & Grubbs, J. B. (2017). The Sexhavior Cycle: Good review, but still not enough data to support a new theory. *Archives of Sexual Behaviour, 46*(8), 2265–2267. https://doi.org/10.1007/s10508-017-1067-5

Love, T., Laier, C., Brand, M., Hatch, L., & Hajela, R. (2015). Neuroscience of Internet pornography addiction: A review and update. *Behavioural Sciences, 5*(3), 388–433. https://doi.org/10.3390/bs5030388

MacDonald, K. (2016, July 21). Signs and symptoms of a porn addiction. *New Zealand Herald*. https://www.nzherald.co.nz/lifestyle/news/article.cfm?c_id=6&objectid=11678076

McKee, A. (2007a). The positive and negative effects of pornography as attributed by consumers. *Australian Journal of Communication, 34*(1), 87–104.

McKee, A. (2007b). "Saying you've been at dad's porn book is part of growing up": Youth, pornography and education. *Metro Magazine*. (155), 118–122.

McKenna, K. (2019, October 11). Kmart kidnapper jailed for 8 years for luring 7yo girl from toy aisle and molesting her in bushland. *ABC News*. https://www.abc.net.au/news/2019-10-11/kmart-abduction-sterling-freesentence/11592880

Mihordin, R. (2012). Behavioural addiction—quo vadis? *The Journal of Nervous and Mental Disease, 200*(6), 489–491. https://doi.org/10.1097/nmd.0b013e318257c503

Montgomery-Graham, S., Kohut, T., Fisher, W., & Campbell, L. (2015). How the popular media rushes to judgment about pornography and relationships while research lags behind. *The Canadian Journal of Human Sexuality, 24*(3), 243–256. https://doi.org/10.3138/cjhs.243-a4

Morrison, C. (2023, October 6). Is porn addiction real? How to tell if you have it and what to do about it. *The Daily Mail.* https://www.dailymail.co.uk/health/article-12598915/porn-addiction-real-statistics-symptoms.html

Moser, C. (2011). Hypersexual disorder: Just more muddled thinking. *Archives of Sexual Behaviour, 40*(2), 227–229. https://doi.org/10.1007/s10508-010-9690-4

Nikunen, K. (2007). Cosmo girls talk: Blurring boundaries of porn and sex. *Pornification, Sex and Sexuality in Media Culture*, 73–87.

Office of Film and Literature Classification. (2018). NZ Youth and Porn: Research findings of a survey on how and why young New Zealanders view online pornography. Office of Film and Literature Classification, Wellington, NZ.

Paasonen, S. (2007). Strange bedfellows: Pornography, affect and feminist reading. *Feminist Theory, 8*(1), 43–57. https://doi.org/10.1177/1464700107074195

Parvez, Z. F. (2006). The labour of pleasure: How perceptions of emotional labour impact women's enjoyment of pornography. *Gender & Society, 20*(5), 605–631. https://doi.org/10.1177/0891243206291109

Plummer, K. (1995). *Telling sexual stories: Power, change and social worlds.* Routledge.

Prause, N. (2019). Porn is for masturbation. *Archives of Sexual Behaviour, 48*(8), 2271–2277. https://doi.org/10.1007/s10508-019-1397-6

Prause, N., Steele, V. R., Staley, C., Sabatinelli, D., & Hajcak, G. (2015). Modulation of late positive potentials by sexual images in problem users and controls inconsistent with 'porn addiction'. *Biological Psychology, 109*, 192–199. https://doi.org/10.1016/j.biopsycho.2015.06.005

Reay, B., Attwood, N., & Gooder, C. (2013). Inventing sex: The short history of sex addiction. *Sexuality & Culture, 17*, 1–19. https://doi.org/10.1007/s12119-012-9136-3

Reinarman, C. (2005). Addiction as accomplishment: The discursive construction of disease. *Addiction Research & Theory, 13*(4), 307–320. https://doi.org/10.1080/16066350500077728

Room, R. (2014). The cultural framing of addiction. In R. Granfield & C. Reinarman (Eds.), *Expanding addiction: Critical essays* (pp. 43–49). Routledge.

Schmitz, J. M. (2005). The interface between impulse-control disorders and addictions: are pleasure pathway responses shared neurobiological substrates? *Sexual Addiction & Compulsivity, 12*(2–3), 149–168. https://doi.org/10.1080/10720160500203641

Skinner, K. B. (2014, November 3). Is pornography (sexual) addiction real? *Psychology Today.* https://www.psychologytoday.com/blog/insidepornaddiction/201411/ispornography-sexual-addiction-real

Sniewski, L., Farvid, P., & Carter, P. (2018). The assessment and treatment of adult heterosexual men with self-perceived problematic pornography use: A review. *Addictive Behaviours, 77*, 217–224. https://doi.org/10.1016/j.addbeh.2017.10.010

Spišák, S. (2016). "Everywhere they say that it's harmful but they don't say how, so I'm asking here": Young people, pornography and negotiations with notions of risk and harm. *Sex Education, 16*(2), 130–142. https://doi.org/10.1080/14681811.2015.1080158

Taylor, K. (2019). Pornography addiction: The fabrication of a transient sexual disease. *History of the Human Sciences, 32*(5), 56–83. https://doi.org/10.1177/0952695119854624

Taylor, K. (2020). Nosology and metaphor: How pornography viewers make sense of pornography addiction. *Sexualities, 23*(4), 609–629. https://doi.org/10.1177/1363460719842136

Taylor, K. (2023). The social diagnoses of digital addictions: Technophobic ambivalences, the limits of the natural and imperatives of self-governance in the information age. *Sociology of Health & Illness.* (Online first) https://doi.org/10.1111/1467-9566.13624

Taylor, K., & Gavey, N. (2020). Pornography addiction and the perimeters of acceptable pornography viewing. *Sexualities, 23*(5–6), 876–897. https://doi.org/10.1177/1363460719861826

Voon, V., Mole, T. B., Banca, P., Porter, L., Morris, L., Mitchell, S., Lapa, T. R., Karr, J., Brand, M., Snagowski, J., Laier, C., & Maderwald, S. (2014). Neural correlates of sexual cue reactivity in individuals with and without compulsive sexual behaviours. *PloS One, 9*(7), e102419. https://doi.org/10.1371/journal.pone.0102419

Vörös, F. (2009). The invention of addiction to pornography. *Sexologies, 18*(4), 243–246. https://doi.org/10.1016/j.sexol.2009.09.007

Walters, G., & Gilbert, A. (2000). Defining addiction: Contrasting views of clients and experts. *Addiction Research, 8*(3), 211–220. https://doi.org/10.3109/16066350009004421

Walton, M. T., Cantor, J. M., Bhullar, N., & Lykins, A. D. (2017). Hypersexuality: A critical review and introduction to the "sexhaviour cycle". *Archives of Sexual Behaviour, 46*(8), 2231–2251. https://doi.org/10.1007/s10508-017-0991-8

Wartella, E., & Reeves, B. (1985). Historical trends in research on children and the media: 1900–1960. *Journal of Communication, 35*(2), 118–133.

Watson, E. (2014, December 9). Pornography addiction among men is on the rise. *Huff Post*. https://www.huffpost.com/entry/pornography-addiction-amo_b_5963460

Weir, K. (2014). Is pornography addictive? *Monitor on Psychology, 45*, 46. https://doi.org/10.1037/e508692014-016

Wendling, M. (2016, February 29). Is "porn addiction" a real thing? *BBC*. https://www.bbc.com/news/blogs-trending-35651737

Wéry, A., & Billieux, J. (2017). Problematic cybersex: Conceptualisation, assessment, and treatment. *Addictive Behaviours, 64*, 238–246. https://doi.org/10.1016/j.addbeh.2015.11.007

Wéry, A., Schimmenti, A., Karila, L., & Billieux, J. (2019). Where the mind cannot dare: A case of addictive use of online pornography and its relationship with childhood trauma. *Journal of Sex & Marital Therapy, 45*(2), 114–127. https://doi.org/10.1080/0092623x.2018.1488324

Wilson, W. C. (1973). Pornography—Emergence of a social issue for study. *Journal of Social Issues, 29*(3), 7–17. https://doi.org/10.1111/j.1540-4560.1973.tb00085.x

Withnall, A. (2013). Pornography addiction leads to same brain activity as alcoholism or drug abuse, study shows. *The Independent*. https://www.independent.co.uk/life-style/health-and-families/health-news/pornography-addiction-leads-to-same-brain-activity-as-alcoholism-or-drug-abuse-study-shows-8832708.html

Zola, I. K. (1972). Medicine as an institution of social control. *The Sociological Review, 20*(4), 487–504. https://doi.org/10.1111/j.1467-954x.1972.tb00220.x

10

Contesting Progress Narratives of Depathologisation in Trans Health

Rebecca Howe

Introduction

For decades, particular gender identities and expressions have been understood in mental illness terms. In a trans health context, it means that psychiatric classifications such as "Transsexualism", "Gender Identity Disorder", "Autogynephilia", and "Gender Dysphoria" have been constructed so that gender self-designation becomes identifiable and thus treatable as a medical problem. Despite significant reversals in recent years, this pathologisation continues to have far-reaching effects. External designations combine with diagnostic assessments in procedural texts to produce a powerful mechanism of governing gender that revolves around approving and providing requested gender-affirming medical technologies. Until recently, international Standards of Care (SOC) required external verification by a mental health professional that a person is

R. Howe (✉)
The University of Melbourne, Melbourne, VIC, Australia
e-mail: rebecca.howe@unimelb.edu.au

N. A. Hendry, E. Karageorgos (eds.), *Critical Mental Health in Australia and Aotearoa New Zealand*, The Politics of Mental Health and Illness,
https://doi.org/10.1007/978-3-031-82861-4_10

psychiatrically distressed-yet-safe-enough to access most medical gender affirmations. This history of incremental change in successive SOCs continues to shape healthcare relations (Lane, 2018). Yet, for generations, people have used multiple means to counter knowledge disqualification and resist the "disordered" classification of gender self-designations that differ from an external attribution at (or before) birth (Davy & Toze, 2018; Hale, 1997; Namaste, 2000; Stone, 1991; Stryker, 1994; Winters, 2022; Worley, 2011). Movements for depathologisation in trans health stem from this sustained, polyvocal opposition to the imposition of psychiatric classifications on people who request access to gender-affirming medical technologies.

Movements for trans depathologisation have largely concentrated on fine-tuning the response to requests for body modification, in which health professionals verify appropriate levels of gender-related distress and capacity for consent. This approach fits a reliably repeated history of bitterly fought battles with a presiding psychiatric authority to create the stand-alone field of trans health and the hard-won improvements in access to gender-affirming medical care (Hanssmann, 2016). It is not to say that a progress narrative is groundless. The unprecedented official and proactive consultation with trans people about depathologising care appears to have generated considerable positive transformation in the development and content of the most recent SOC-8 (Coleman et al., 2022; Malatino, 2020). Driven by trans activism and scholarship, part of this change results from a demand to shift from an illness-based model to a rights-based approach (Kara, 2017; Schwend, 2020; Schwend et al., 2014). Yet concerns remain about an approach to depathologisation that maintains the authority of psychiatric knowledge in determining who is granted access to care (e.g., Winters, 2022).

A linear story of gradual improvement leaves little room to discuss how activist demands are also contained and co-opted to preserve, legitimate, and expand an existing care system. Nevertheless, critical mental health scholars from a range of disciplines in Australia and Aotearoa New Zealand, regularly interrogate such an appealing view of medicalised approaches to suicide prevention (Fitzpatrick & River, 2018), the concept of pornography addiction (Taylor, 2019), and mental health awareness-raising campaigns (Tseris, 2019). I join the contestation of a progress narrative because it limits the full extent of depathologisation. Dissensus is less about abandoning notions of progress in trans health than broadening the account of trans depathologisation. This analysis is part of a continued contestation of care itself, aligning with the aims of political movements against pathologisation by defamiliarising underpinning logics to challenge an assumed fixed status of its authority. Accordingly, this chapter examines how requesting access to gender-affirming medical technologies *came to be* represented as a mental health problem requiring psychiatric assessment and diagnosis to enable "treatment" of the body.

Genealogy provides a valuable framework for analysing this pathologisation as a form of governing gender *through* governing access to gender-affirming medical technologies. Foucault and Burchell (2008) famously understood governing as emerging from within "a dense web" of power relations that transit "through apparatuses and institutions" and a "swarm of points of resistance" (p. 96). Yet critiques from Indigenous studies and Black studies scholars generatively challenge Foucault's inadequate attention to colonisation and racialisation as interconnected historical phenomena that direct the course of action and struggle (e.g., Moreton-Robinson, 2015; Weheliye, 2014). Consequently, this chapter outlines some of the current conditions of trans health before considering eight critical moments in care across *time* while attending to the significance of *place*. The following genealogical account of how people with self-designated genders have become psychiatrised subjects of care helps to surface its liberal assumptions and account for distinct but connected colonial capitalist foundations. This critical analysis of the emergence of conditions, their ongoing contestation and possibilities counter a limiting progress narrative of depathologisation in trans health. Working from an established critique of an illness-based approach, this analysis also supports a critical evaluation of an ascendant model of rights as an emancipatory alternative paradigm for care in trans health. Attending to unexamined logics and inviting complexity broadens the understanding of the necessity and scope of such work.

The Conditions of Trans Healthcare in the Present

In a Foucauldian sense, genealogy is an analytic that starts from a concern in the present. Tamboukou (1999) explains how a present moment is theorised as an *episode* within an ongoing struggle to highlight the importance of considering time and place. This genealogy of care starts from the present objection to the continued pathologisation of people who request access to gender-affirming medical technologies. The concern about pathologisation is shaped by an Australian research context that details continuing, multilayered experiences of discrimination, stigma, marginalisation, and harm in relation to non-normative gender identity and expression (Callander et al., 2019; Hill et al., 2020). These are not just interpersonal but structural harms enacted within healthcare and an array of other services, systems, and policies. Disputes over determining access to care have generated alternative visions for depathologisation that redesign healthcare systems to meet rather than regulate requests for body modifications. These alternatives are not simply in response to the ways pathologisation structures harmful practices and inflexible relations but also a growing appreciation of how it operates as part of a mechanism that helps secure oppressive colonial state power. Medical technologies, procedural texts, and external verification are three key components of this mechanism that create the material conditions of trans health.

Medical technologies mediate the conditions of trans health. While medical technologies describe the means of body modification, affirming refers, more broadly, to particular ways a person may move across a socially imposed gender boundary (Stryker, 2017). Different eligibility criteria and assessments determine access to different medical technologies depending on which combination of hormone or surgical technologies a person has requested. Assessment and diagnosis are as ubiquitous in trans health as they are in other healthcare settings. With gender-affirming medical technologies typically framed as "treatments" for mental distress, psychiatric diagnoses preserve patterns of power and authority for granting access—not only care but also identity documentation and, through this, access to rights of citizenship. At this broad level, Johnson

(2015) contends that a medical logic that simultaneously legitimises and restricts access to gender-affirming medical technologies has developed into "a hegemonic medical model for understanding transgender people's experiences of gender over the life course" (p. 804).

The relations between healthcare providers and those requesting care are thus operationalised via diagnoses in procedural texts. Following Burstow's accounting (2015), these are understood as governmental boss texts that include the Diagnostic and Statistical Manual of Mental Disorders (DSM), along with the International Classification of Diseases (ICD) and the Standards of Care (SOC). Attention to the politics of external verification shows how recognition functions in medical contexts to acknowledge the existence, legitimacy, and legality of being. Ansara and Hegarty (2014) argue that because "authorities in English language contexts typically assign 'permanent' gender at or before birth ... people's own descriptions of their genders and bodies are treated as invalid unless they are typically associated with their assigned sex designations" (p. 259). Requirements of external verification of a person's description of their gender, distress, and capacity by healthcare providers configure a hierarchy of status relations within and beyond Western medicine. Sustained objections from trans depathologisation activists to the ways mental illness diagnoses are used to regulate access to care connect to broader struggles over the ways that gender-affirming medical technologies are tethered to multiple forms of social and state recognition (Aizura, 2006). The following analysis of the emergence of conditions and the contestation (thus possibilities) in care is shaped by the provision of medical technologies, the creation of authoritative procedural texts, and proposals for intervention into mechanisms of external verification entrenching a regime of status relations.

Emergence, Contestation, and Possibilities in Care

The concern at the centre of this genealogy of care is how requesting access to gender-affirming medical technologies has come to be tied to pathologisation. This section outlines eight critical moments when this linking and (attempted) de-linking occur to re-consider the prevailing progress narrative that shapes trans health policy. As Tamboukou (1999) explains, genealogies do not theorise the present "as the result of a meaningful development, but rather as an episode, a result of struggle and relations of force and domination" (p. 203). As Amin (2018) argues, a premise of "multiple, contingent origins rather than a singular, determining one" helps understand "historical time not as a causal, straight line but as a fragmented and tangled network of ruptures, continuities, reversals, and reconfigurations" (p. 602). Consequently, each critical moment was selected for the unique way it reveals a kind of rupture, continuity, or a (re)configuration of care. A genealogical approach usefully expands the consideration of conditions shaping the depathologisation of care in trans health.

This analytic method includes a profusion of moments in various places. Consideration requires attention to practices, struggles, and relations within, alongside, and seemingly far away from trans health. The chosen moments reveal the emergence, coalescence, and contestation of pathologisation, or what Dewey and Gesbeck (2017) identify as "the diagnostic processes and categories that legitimate gender in the medical system" (p. 39). Yet, these moments also reveal important overlaps and connections by extending across and beyond medical systems. These moments are critical because they help clarify how "care" secures, authorises, and protects hegemonic medical authority and colonial state power in multiple places. This clarification directs attention to the ways trans depathologisation efforts respond to different tactics for governing gender and governing access to care. Thus, in calling attention to contingent governing, this genealogy traces the conditions and contestation of care in trans health.

1930s Production of Synthetic Hormones

Turning from pathologisation in the present to the production of synthetic hormones in the late 1930s reveals an important history of medical technologies. Synthetic manufacturing was made possible, in part, by pharmaceutical companies supplying research teams with the hormone-rich urine of pregnant horses purchased from farmers and slaughterhouses, the urine of pregnant women obtained from gynaecological clinics, and the urine of men collected from factories, military barracks, and prisons across Germany and Holland (Oudshoorn, 2003). While these urine preparations played a central role in studying hormones, they also had earlier connections to organotherapy. This treatment modality used extracts from animal glandular organs to restore a youthful appearance, increase virility, and boost masculinity (Rubin, 2006). Organotherapy has enjoyed some popularity since at least the 1890s (Oudshoorn, 2003), and by the 1920s, transplanting parts of chimpanzee testes into elite European men had gained widespread attention (Rubin, 2006). Scientific communities, the press, and even speculative fiction writers at the time responded with outcry at "the possibilities of human-ape ontological and erotic proximity" (Amin, 2020, p. 49) or simply the perceived implications of inter-species mixings (Rubin, 2006). The accompanying practice of organ extraction, however, received far less critical attention than organ implantation.

Material for many of the early "rejuvenating" experiments was taken from chimpanzees amid waves of French colonial genocide in Algeria, and both extractive endeavours shared the same colonial eugenic premise of maintaining the position of "the West's ideal man" (Rubin, 2006, p. 486). However, as the promise of animal extracts improving sexual performance went unfulfilled, these practices were increasingly dismissed as medical quackery (Oudshoorn, 2003; Rubin, 2006). However, this preoccupation with "the unique potencies" of genitals persisted as a form of eugenic thought in the emerging field of endocrinology that "envisioned sex as at once the sign of racial (un)fitness and the easily manipulable means of improving the [white] race" (Amin, 2018, p. 601). As endocrinology and its distinction between a person's sex (morphological and physiological), sexuality, and gender were legitimised, people with self-designated genders who requested body modifications could now become people with bodies that were "treatable" with testosterone or oestrogen and progesterone (Rubin, 2006).

Research using hormone-rich bodily fluids paved the way for synthetic hormone production. With pharmaceutical companies now able to provide any quantity of synthetic hormones at a reduced cost, new uses for hormone technologies increased exponentially (Oudshoorn, 2003). Crucially, these advancements helped establish hormones disciplining applications in psychiatric and other carceral institutions (e.g., Ryan, 2022 in a US context). While synthetic hormone manufacturing may have interrupted one form of colonial, white supremacist eugenic thought and practice, these were reincorporated into care through producing, imposing, and regulating access to synthetic hormone technologies. This was a transatlantic configuration of medical technologies extracting across Europe and North Africa and extending into the US. The German-American endocrinologist and sexologist Harry Benjamin was an early adherent and practitioner of organotherapy, attaining celebrity doctor status after decades of "revitalising" men and "reactivating" women in the US (Amin, 2018). Benjamin played a prominent role in influencing endocrinology's separation of sex, sexuality, and gender into categories that still organise dominant understandings of bodies and identities. Benjamin's distinguished position also meant the reach of these categories extended beyond endocrinology to surgery and the psy-professions, influencing the creation of rigid criteria for which technologies can be used as "treatments" for particular bodies.

2021 Addition of Voice Feminisation Surgery to Israel's State-Funded Services

Another moment in the present shows how Israel incorporates a discourse of a liveable life for trans people into a particular localised political, cultural, and social context of an established state-funded healthcare system. In August 2021, it was announced that voice feminisation surgery would be added to Israel's state-funded services. On the surface, this news appears as a critical reconfiguration of care, considering that calls for state-funded coverage have long been a central component of an international, human rights-focused trans depathologisation movement (Schwend, 2020). Health Minister Nitzan Horowitz framed the decision as the response of a state that has, for too long, "pushed many members of the [trans] community to the margins of society, to a life in difficult conditions and to the street" (Joffre, 2021, para. 1). In response to high rates of suicide attempts, reports of discrimination and refusals from practitioners to provide care, Minister Horowitz argued that access "isn't just another ordinary struggle, it is a struggle to be who you are, a struggle for freedom, a struggle for life", and "as elected officials and as a state our moral duty is to allow every person to live their lives as they choose" (Joffre, 2021, para. 11). Yet this discourse of Israeli state-funded healthcare allowing a liveable life—particularly for trans women—circulates amid estimations that only 20 people access gender-affirming surgeries in Israel each year, with an even smaller number seeking out surgical "treatments for gender adjustment of the voice" (Spiro, 2022, para. 2). These surgeries to change the tension of a person's vocal cords effectively amount to less than half a per cent of the 550 million Israeli Shekel (nearly 250 million Australian dollars) allocated for all newly approved procedures and medications (Spiro, 2022). Israel appears to address the demand for state funding of gender-affirming medical procedures by rationalising it as a life-saving response to the ill-health effects of violence, mental distress, family and social exclusion, employment, and housing discrimination (Joffre, 2021; Spiro, 2022). The point is not to question the necessity of funding surgeries. Rather, it is to highlight how this justification for state subsidy operates at a distance from the provision of care as part of a responsibility to uphold a broad range of rights and to recognise that trans lives are lived in multiple ways that precede and exceed a request for body modification.

Israel's mobilisation of a neoliberal-inflected discourse of an individualistic response to experiences of harm reinforces an illness paradigm as it attempts to recuperate the state from marginalising (denying) to allowing (saving) the lives of people with self-designated genders. Moreover, it demonstrates how activist demands for access to gender-affirming medical technologies in a state-funded health system are repurposed as a component of an extensive, well-documented pinkwashing project of Israel presenting itself as a progressive, welcoming, LGBT rights-upholding state to justify its ongoing illegal settler colonial invasion and genocidal occupation of Palestine (e.g., Puar, 2017). Scrutinising how Israel subordinates the public funding of gender-affirming medical technologies to the prevailing illness-based model of care reveals how easily the discursive and literal investments in trans health are folded into the state-securing logics of settler colonialism. This localised critical moment of a superficial reconfiguration has international significance in that it illuminates a recurring pattern of transformations that affect just enough change to give the appearance of ending harmful practices while preserving the dominant status regime (Siegel, 1997; Spade, 2013).

2016 Gender-Affirming Genital Surgery Skills Development Workshop

Traversing five years in another direction is a surgery skills development workshop held in 2016 at an inner-city hospital in the Midwest United States. This critical moment of teaching and advocacy in a hospital setting compels further attention to the conditions required to realise material access to medical technologies. Plemons (2018) describes how twenty surgeons participated in a two-day lecture and cadaver lab demonstration of metoidioplasty (creating a penis and scrotum from existing genital tissue) and phalloplasty (constructing a "neophallus" using skin grafts from the person's forearm or thigh). The World Professional Association of Transgender Health (WPATH) organised the workshop in response to the increasing demand for and interest in providing genital surgeries (Plemons, 2018). This growth is partly due to a 2014 lawsuit that effectively ended the exclusion of these surgeries in the US Medicare system, combined with fifteen years of pushing private, employer-provided insurance to increase coverage for gender-affirming medical care (Plemons, 2018). Guaranteed payments for genital surgeries were an enticing income source that increased hospital-based teaching programs. Yet, this possibility for more affordable options also encroached upon a small, competitive, international market that caters to those with resources to pay directly for care (Plemons, 2018). The workshop also sought to address a situation where most surgeons within the US market were reticent to share their lucrative technical expertise (Plemons, 2018).

Crucially, becoming proficient in complex genital surgeries cannot occur in a workshop setting. Skills development requires specialised clinics where surgeons can practise "operating on particular tissues and structures", foster "deep anatomical knowledge of the area of concern", and generate "technical fluency of particular procedures" (Plemons, 2018, p. 6).

Further difficulties in realising access to genital surgeries occur because the SOC had essentially prescribed "the cooperation of distinctly credentialed specialists" across "multiple stages of authorization and permission" (p. 10) in ways that create "layered administrative responsibilities" (Plemons, 2018, p. 13). Here, the emphasis is not on how people with self-designated genders exceed the limits of current hospital and insurance systems; it is on how surgical requirements such as those set out in the SOC make a complicated situation more difficult.

Considering the specificities of surgical procedures nuances this critical moment of possibility in care. Especially as the term "gender-affirming surgery" can obscure how many of these surgeries exist and that they "happen in phases, require revisions, and are often the result of several individual procedures combined" (Rosenberg et al., 2021, p. 1). For Plemons (2018), these challenges of bureaucratising care "demonstrate the tangled and peculiar ways that trans-medicine has been imagined outside of these kinds of technocratic institutions" (p. 13). The messiness certainly indicates a late liberal imagining of including trans health into an existing US economic-health system. There is also an important moment of rupture, illustrated in the workshop organisers admitting that although the exercise "pulls back the shroud" on otherwise hidden procedural knowledge, it only provides familiarity and does not equip workshop participants with the necessary skills to carry out these surgeries (Plemons, 2018, p. 8). The surgical skills development workshop provides a window into the operation of a tightly regulated, economically oriented care mechanism with fixed and moving parts. The workshop compels attention not only to institutional and technical practices but also to the vigilant and radical imagination required to transform obtaining permission into the use of medical technologies.

1979 Johns Hopkins Gender Identity Clinic Closure and the First SOC

Turning to a different time but staying in the eastern part of the US, the next critical moment looks at the 1979 Johns Hopkins Gender Identity Clinic closure and the creation of the first SOC. Despite funding concerns, communication difficulties, and staffing issues, Siotos et al. (2018) show how the Clinic closure was the culmination of public outcry over a published evaluation of the program and the vitriolic antitransgender stance of Dr Paul McHugh, head of Psychiatry at the Hospital. Crucially, the much-publicised debate on "medical morality" and gender-affirming surgery that helped end surgeries at Johns Hopkins (Siotos et al., 2018, p. 2) also helped justify the closure of almost all programs offering gender-affirming medical care in US academic-teaching hospitals which coincided with the elimination of even the minimal health insurance coverage for gender-affirming medical care (Gorton & Grubb, 2014). Indeed, as Stryker (2017) argues, this erosion of access, particularly under the Reagan administration in the 1980s, was partly shaped by "antitransgender feminist arguments that dovetailed with conservative politics" (p. 112). Highlighting the lessons of this critical moment is crucial, considering how trans-led scrutiny into the evidence-gathering methods of contemporary campaigns blocking access to care is ongoing in the US and UK (e.g., Ashley et al., 2023; Trans Safety Network, 2024).

The 1979 Clinic closure coincided with the publication of the first SOC by the newly established Harry Benjamin International Gender Dysphoria Association (HBIGDA), which would later become the World Professional Association of Transgender Health (WPATH) (Coleman et al., 2012). A perceived lack of institutional support created a seemingly heavy responsibility for medical professionals and a certain intensity of feeling about the impact of media controversy and public backlash on their status and livelihoods (Stryker, 2017). HBIGDA provided relief by formalising the position of psy-professionals as the first in what Serano (2016) describes as a series of gatekeepers whose role was to regulate those obtaining gender-affirming medical technologies. Establishing this external verification process as a standard is significant; access to care was sustained by focusing on regulating rather than increasing its provision. As Velocci (2021) makes clear, prominent practitioners like Harry Benjamin "did not define transness by establishing solid criteria and then assessing people by how closely they matched them; rather, they constructed transsexuality as a space of uncertainty through their fear-based practices and used that uncertainty to maintain clinical control" (p. 464). Despite a steadfast declaration that the early Standards were a safeguard against unsatisfactory clinical outcomes, Serano (2016) argues that it inculcated practices to limit access and, further, to isolate and invisibilise those allowed through the gateway. The development of the first SOC suggests a political process of installing an eligibility standard *as* a care standard that continues as authorisation and protection for health professionals as regulators of access to gender-affirming medical technologies. This international policy document has undergone eight revisions over forty-three years. Yet even while adopting a rights-based approach, the SOC continues to contend with the legacy and assumed worldwide application of this foundational attribution of authority to medical knowledges and judgement by expert health professionals when responding to requests for access to gender-affirming medical technologies.

1893 Creation of the ICD

Another critical moment of configuration, but from the previous century, is the 1893 establishment of the ICD. The first iteration of the ICD was a codification of specific causes of death developed by Jacques Bertillon, presented at a meeting of the International Statistical Institute in Chicago in 1893, promoted by the French International Office for Public Hygiene, and implemented by the statistical agencies of multiple governments over the next thirty years (Bowker, 1996). With eleven subsequent iterations, ICD codes have come to represent diagnoses and health status determined through healthcare procedures (World Health Organisation, 2020). On the surface, this critical moment may seem far from care in trans health. However, considering how counting bodies according to disease classifications reveals how care is also configured as surveillance for managing populations, further illuminating the capitalist colonial conditions that produce some bodies as countable and others as "treatable".

The spread of epidemics after intensive European colonisation provided an imperative for codifying causes of death. Bashford (2003) notes that Muslims on pilgrimage from India to Saudi Arabia were a key factor in a major cholera outbreak across Europe. Steamboat and rail technologies, driven by the colonial economic imperatives to open the Suez Canal, increased the spread of cholera to pandemic proportions (Bashford, 2003; Bowker, 1996). Pilgrims now arrived home from Hajj while unwell, increasing transmission, whereas previously, those who returned on foot often died in the time it took to travel (Bowker, 1996). Concerns that cholera outbreaks now impacted citizens of colonising rather than colonised nations resulted in a confluence of state and commercial practices regulating infectious disease transmission, migration, and movement (Bashford, 2003). These practices generated increasingly racialised and eugenic distinctions between populations subject to different forms of inspecting, banning, quarantining, and confining under rationales of health and "hygiene" (Bashford, 2003). Regarding the ICD, Bowker (1996) describes how the French Empire imposed its classification of causes of death on its colonies, which other imperial powers reproduced. While these efforts were regularly contested and rejected, this scheme led to the development of modern bureaucratic state structures to monitor the ill health and death of populations on one hand and the justification for population-level surveillance methods on another (Bowker, 1996).

Over time, the enormous amount of information needed to operationalise a system of medical coding involved developing uniform bureaucratic practices that corresponded with the classification system. This necessitated habituating people to regular collection of information about ever-increasing aspects of their lives, standardising these procedures to produce comparable data, and developing a class of educated professionals to use this information (Bowker, 1996). These are the structures, services, systems, and policies from which multiple forms of harm are enacted against people with genders and bodily configurations that do not conform to normative categories. Crucially, "building the ICD involved building the state as much as developing medical knowledge" (Bowker, 1996, p. 52). Thus, gender and gendering processes are interwoven into these different scales of governing. Revealing histories of configuring care to surveil populations for governing also reveals it as unfinished work, suggesting the potential for more radical, anti-colonial unmaking of this "care". Trans depathologisation activism and transgender studies sustain their pivotal role in this work at different scales in different contexts by attending to these colonial conditions (e.g., Boellstorff et al., 2014; Maggic, 2017; Stanley & Smith, 2016).

2012 Proposed Starfish Model of Codes Classification for ICD

Arriving at another relatively recent moment shows that attempts to radically reorganise this medical coding system occur through engaging with complexity and disagreement. In November 2012, Global Action for Trans Equality (GATE) organised a meeting to discuss the imminent review of the ICD-10. The meeting was convened because "questions raised by the need to combine depathologization and full access to health and legal rights do not have a single answer that can satisfy the needs and interests of all trans* persons in every country and public or private health system" (GATE, 2012, p. 3). The meeting, aimed to suggest "lines of work" for the remaining two years in the development of ICD-11, was attended by health and other experts from fourteen different nations, most of whom identified as trans (GATE, 2012).

The ICD comprises chapters containing blocks (that group-related references), codes, and diagnoses. An idea for an alternative, distributed system emerged from a position that ICD references (any chapters, blocks, or codes) must not be illness-based yet be capable of facilitating access to healthcare, insurance coverage, and legal recognition (GATE, 2012). Codes must be broad enough to sustain access to care as it changes over time but not too wide that they miss attention to a person's needs (Eisfeld, 2014, p. 108). A metaphoric Starfish Model was used to describe this idea because these marine invertebrates have a decentralised nervous system. "While all parts of the starfish work together, a chopped-off leg can exist on its own, and actually grow a new starfish" (GATE, 2012, p. 10). In contrast, a spider was used to describe the existing model that "has a centralized nervous system where the whole spider dies if the head is injured" (GATE, 2012, p. 10). Applying the spider and the starfish metaphor articulates how "combining different [ICD] chapters, blocks and codes (the starfish's legs) could exponentially increase trans* people opportunities of accessing health care under very different circumstances without recurring to a single and potentially re-pathologizing diagnosis (a Spider Model)" (GATE, 2012, p. 10).

The transformative potential of the Starfish Model appears in the meticulous attention given to addressing complex layers of pathologisation in the current systems of care in trans health. The meeting reached agreements on removing ten codes from the ICD that maintained stereotyping, pathologising, stigmatising, and gatekeeping practices and justified continuing human rights abuses (GATE, 2012). Including rights abuses in places "where excessive and irrelevant testing in psychiatric hospitals is required for any change of ID documents" (GATE, 2012, p. 6). While there was consensus on matters of code removal, disagreement on issues of diagnosis as a need, the use of a decentralised model for change, determining the placement of chapters, blocks, codes and diagnoses, and how to support children and young people could not be resolved (GATE, 2012). Consequently, the meeting report offers six recommendations for developing the ICD while documenting the supporting arguments, challenges, and additional comments to promote further discussion (GATE, 2012). The difficulty of operationalising a decentralised model for the ICD reflects the intricacies of listing and classifying all the illnesses, injuries, and health-related situations that require diagnosis, verification, and processing in a health system. The unresolvable issues raise further questions of whether trans depathologisation movements are reorganising or dismantling the ICD as part of the mechanism of care. While the Starfish Model was not put forward as a proposal, this attempt at reorganising was not a dead-end. This expert meeting exemplifies a sustained ethico-political commitment to trans depathologisation. Many ICD-11 changes occurred due to this active involvement and recognition of people with self-designated genders' expertise in articulating and (re)structuring responses to their healthcare needs.

1789 Declaration of the Rights of Man and Citizen

Rebounding across several centuries, the 1789 Declaration of the Rights of Man and Citizen is a different example of group decision-making. This time, an unwieldy committee during the French Revolution attempting to draft and approve an official declaration of principles for governing. This declaration is a critical moment of political expression whose implications reveal the importance of examining how human rights may be used as an alternative framework for care in trans health. Hunt (2007) describes how the National Assembly, which declared themselves representatives of the French people in the newly formed French Republic, needed to go beyond writing speeches and making laws. In writing, the Assembly articulated the pre-existing rights that were the basis of their new government. The subsequent Declaration proclaims, "Men are born and remain free and equal in rights", and as Hunt (2007) clarifies, these are "not French men, not white men, not Catholics, but 'men,' which then as now means not just males but also persons, that is, members of the human race" (p. 23). A universalising but explicitly gendered and racialising language of rights and belonging that emerged during this political struggle in the eighteenth century helped crystallise these notions as unquestionable, prompting excluded groups to make new demands for rights (Hunt, 2007). Drawing on the historical literature, Hunt (2007) outlines how granting and refusing the rights of Protestants and Jews as religious minorities, enslaved and racially marginalised people, and women were in constant flux over subsequent decades. Locating the Declaration historically and geographically highlights how these notions of human rights were inextricable from questions of granting citizenship.

How rights were declared and recognised in France compared to its colonies highlights the importance of geopolitics in officially sanctioned state membership. Haiti is one example of the implications of this articulation of universal rights. After initially excluding all its colonies from the Declaration, France strategically voted to abolish slavery during the Haitian Revolution in 1794 in an attempt to maintain some control of an island that generated immense wealth not only for the French but also for British and Spanish colonial powers and the US (Alcenat, 2021; Hunt, 2007). In stark contrast to the Haitian revolutionaries, France moved to emancipate enslaved people and bestow citizenship rights by assuming the complete elimination of the prior and continuing presence of the Taíno (Alcenat, 2021). For Smith (2021), it is the tenet of "humanity" itself that functions within established, gendered, and hierarchical "systems of rule and forms of social relations which governed interaction with the Indigenous peoples being colonized", making it possible for colonisers "to consider Indigenous peoples as not fully human, or not human at all", in ways that "enabled distance to be maintained and justified various policies of either extermination or domestication" (p. 99). Articulating human rights as a basis of governing without transforming established gendered and racialising systems of rule maintains a continuity of relations of supremacy and domination.

While this is another critical moment that seems far removed from care in trans health, the Declaration's historical and geographical reverberations highlight how this revolutionary break from the rule of an absolute monarchy was also a continuity of its colonial capitalist foundations. Of particular concern in the context of trans health, is the continuity of systems of rule that make it possible to operate at a distance to determine someone else's existence and status. This legacy appears in a dominant late liberal framing that bypasses how rights are situated and articulated within a web of governing practices, not least those of establishing and maintaining nations through gendering and racialisation. Yet still, Hunt (2007) argues that "rights remain open to question because our sense of who has rights and what those rights are constantly changes" (p. 29). The further example of Haiti, especially its own Declaration of Independence, draws attention to the existence of alternative and emancipatory ways of governing. Histories of human rights raise an important question about uncritically mobilising (and celebrating) liberal human rights as a framework for care. What happens when an attempt to replace harmful practices and relations does little to address the enduring conditions that structure contemporary violence and harm? Attending to this history reveals that rights are also open to question because of what rights do.

2017 Yogyakarta Plus 10

Transiting two hundred years reveals another contemporary reckoning with recognising and applying human rights. The final critical moment in this genealogy focuses on adding to the Yogyakarta Principles (YPs) (2007) on the Application of International Human Rights Law in Relation to Sexual Orientation, Gender Identity, Gender Expression, and

Sex Characteristics. The YP plus 10 (2017) was created on the tenth anniversary of the YPs and added ten principles at the point where "developments in international human rights law" met "with the emerging understanding of violations suffered by persons on grounds of sexual orientation and gender identity" and combined with "recognition of the distinct and intersectional grounds of gender expression and sex characteristics" (p. 4). Broadly, the Principles presuppose sexuality and gender as universal characteristics (Park, 2019). Taking this position was a way to resolve the disjuncture of describing rights held only by LGBTI people within an enduring discourse that defines human rights as held by *all* humans. As a result, both statements emphasise human characteristics over specific populations. The rationale is that all people have a sexual orientation, a gender identity, gender expression, and sex characteristics (SOGIESC). Likewise, all people can experience rights violations across any of these characteristics. The purpose of the full set of 29 Principles is to clarify a State's obligations as a signatory to human rights conventions to prevent and address human rights violations related to a person's SOGIESC. Carpenter (2020) explains that this advice on fulfilling obligations and applying existing human rights legislation was necessitated by the lack of a comparable UN convention, such as the rights of the child and persons with disabilities.

The YP plus 10 connects to concerns about the pathologisation of people requesting access to gender-affirming medical technologies through the position it takes on how a government should fulfil its obligations as a signatory of human rights conventions. Despite member's misgivings, the 2007 YPs framed human rights violations experienced by people with non-normative relationships, genders, and bodies "in terms of inclusion in current legal norms" without "structural reform of those norms" (Park, 2019, p. 248). Doubts developed into a challenge to the liberalising notion of inclusion (e.g., Otto, 2016; Waites, 2009) and grew into a well-established critique of the inappropriate normative framings of sexual orientation and gender identity and the exclusion of intersex people in the YPs (Park, 2019; Carpenter, 2020). A practice of inclusion was identified as a limitation in the applicability of the YPs, especially in mobilising responses to violations of the human rights of intersex people (Carpenter, 2020).

Consequently, the YP plus 10 emphasises the influence of compounding oppression. That is, explicitly recognising "the intersectionality of sexual orientation, gender identity, gender expression and sex characteristics, including on grounds of sex, gender, racialisation, religion, belief and disability" (Carpenter, 2020, p. 523). This shift in approach potentially increases the capacity for advocates and activists to push for recognition of the structuring violence of the law itself. Nevertheless, when taken together, the Principles are understood as a set of "escalating demands, from inclusion to recognition" through "a consensus that LGBTI people fit within the international human rights regime" (Park, 2019, p. 248). This critical moment is particularly relevant because trans depathologisation activists within the Stop Trans Pathologization (2010) network invoked the YPs in its submission to the SOC-6 revision process that argued for promoting "trans-specific health rights" concurrently with a move to *replace* a paradigm of illness with a paradigm of rights (p. 12). The YP plus 10 and the YPs are thus presented as valuable tools for activists and advocates at both local and international levels. However, the ten additional Principles increase their effectiveness by broadening the "framework beyond matters of self-identification" to the fact of SOGIESC (Carpenter, 2020, p. 528). This example of ongoing commitment signals possibilities for further reconfiguring care not just by replacing the current illness model but also by addressing the entire framework that constitutes the dominant understanding of human rights.

Conclusion

This genealogy has considered eight critical moments of governing gender and governing access to gender-affirming medical technologies. The purpose of this temporal and geopolitical consideration has been to highlight the conditions and contestation from which depathologisation in trans health has emerged. Tracing these conditions of care is important because it reveals how colonial rationales and practices of governing are foundational and enduring. Tracing its contestation also reveals how these conditions are overlaid with different strategies and tactics for governing access to gender-affirming care. All of this is important because it emphasises why critical attention needs to be sustained on the development of trans health policy and where this attention can go.

Understanding how trans people and people with nonbinary genders have become psychiatrised subjects of care within colonial projects is an essential part of dismantling an illness model and critically evaluating the process of replacing it with a human rights model. By examining how hegemonic medical authority and state power are installed, authorised, and secured, we can appreciate how trans depathologisation efforts are undertaken within and often against different techniques of governing access and care. This approach offers a generative challenge to an empty assertion of depathologisation and the maintenance of recuperative, obscuring, and settler state-securing colonial conditions of neoliberal racial capitalism. This understanding is important, particularly for cis people, in considering their political position and ethical stance in undertaking analyses and (re)formulating collective strategies that create the conditions for a liveable life. The mechanisms of politics that allow for trans life are thus entangled within colonial projects, regardless of whether they involve establishing and maintaining a settler state or another colonial formation. It is for this reason that complete depathologisation in trans health, or any setting, cannot occur within contexts of continued colonial dominion and state power. Recognising colonisation as a primary condition of modern governing raises valuable questions about uncritically mobilising a human rights paradigm to replace an illness model for the depathologisation of care.

References

Aizura, A. Z. (2006). Of borders and homes: The imaginary community of (trans) sexual citizenship. *Inter-Asia Cultural Studies, 7*(2), 289–309. https://doi.org/10.1080/14649370600673953

Alcenat, W. (2021). How U.S. economic imperialism underdeveloped Haiti. *NACLA Report on the Americas, 53*(2), 193–201. https://doi.org/10.1080/10714839.2021.1923226

Amin, K. (2018). Glands, eugenics, and rejuvenation in man into woman: A biopolitical genealogy of transsexuality. *Transgender Studies Quarterly, 5*(4), 589–605. https://doi.org/10.1215/23289252-7090059

Amin, K. (2020). Trans* plasticity and the ontology of race and species. *Social Text, 38*(2 (143)), 49–71. https://doi.org/10.1215/01642472-8164740

Ansara, Y. G., & Hegarty, P. (2014). Methodologies of misgendering: Recommendations for reducing cisgenderism in psychological research. *Feminism & Psychology, 24*(2), 259–270. https://doi.org/10.1177/0959353514526217

Ashley, F., Tordoff, D. M., Olson-Kennedy, J., & Restar, A. J. (2023). Randomized-controlled trials are methodologically inappropriate in adolescent transgender healthcare. *International Journal of Transgender Health, 1–12*. https://doi.org/10.1080/26895269.2023.2218357

Bashford, A. (2003). *Imperial hygiene: A critical history of colonialism, nationalism and public health*. Springer.

Boellstorff, T., Cabral, M., Cárdenas, M., Cotten, T., Stanley, E. A., Young, K., & Aizura, A. Z. (2014). Decolonizing transgender: A roundtable discussion. *Transgender Studies Quarterly, 1*(3), 419–439. https://doi.org/10.1215/23289252-2685669

Bowker, G. C. (1996). The history of information infrastructures: The case of the international classification of diseases. *Information Processing & Management, 32*(1), 49–61. https://doi.org/10.1016/0306-4573(95)00049-M

Burstow, B. (2015). Probing the boss text: The DSM—What? Whither? How? Which? In B. Burstow (Ed.), *Psychiatry and the business of madness: An ethical and epistemological accounting* (pp. 73–100). Palgrave Macmillan US. https://doi.org/10.1057/9781137503855_4

Callander, D., Wiggins, J., Rosenberg, S., Cornelisse, V., Duck-Chong, E., Holt, M., Pony, M., Vlahakis, E., MacGibbon, J., & Cook, T. (2019). *The 2018 Australian Trans and Gender Diverse Sexual Health Survey: Report of Findings.* U. S. The Kirby Institute. https://kirby.unsw.edu.au/report/2018-australian-trans-and-gender-diverse-sexual-health-survey-report-findings

Carpenter, M. (2020). Intersex human rights, sexual orientation, gender identity, sex characteristics and the Yogyakarta principles plus 10. *Culture, Health & Sexuality, 23*(4), 516–532. https://doi.org/10.1080/13691058.2020.1781262

Coleman, E., Bockting, W., Botzer, M., Cohen-Kettenis, P., DeCuypere, G., Feldman, J., Fraser, L., Green, J., Knudson, G., & Meyer, W. J. (2012). *Standards of Care for the Health of Transsexual, Transgender, and Gender*

Nonconforming People, 7th Version. The World Professional Association for Transgender Health.
Coleman, E., Radix, A. E., Bouman, W. P., Brown, G. R., de Vries, A. L. C., Deutsch, M. B., Ettner, R., Fraser, L., Goodman, M., Green, J., Hancock, A. B., Johnson, T. W., Karasic, D. H., Knudson, G. A., Leibowitz, S. F., Meyer-Bahlburg, H. F. L., Monstrey, S. J., Motmans, J., Nahata, L., et al. (2022). Standards of care for the health of transgender and gender diverse people, Version 8. *International Journal of Transgender Health, 23*(Suppl. 1), S1–S259. https://doi.org/10.1080/26895269.2022.2100644
Davy, Z., & Toze, M. (2018). What is gender dysphoria? A critical systematic narrative review. *Transgend Health, 3*(1), 159–169. https://doi.org/10.1089/trgh.2018.0014
Dewey, J. M., & Gesbeck, M. M. (2017). (Dys) functional diagnosing: Mental health diagnosis, medicalization, and the making of transgender patients. *Humanity & Society, 41*(1), 37–72. https://doi.org/10.1177/0160597615604651
Eisfeld, J. (2014). International Statistical Classification of Diseases and Related Health Conditions. In *Transgender Studies Quarterly* (Vol. 1, pp. 107–110). https://doi.org/10.1215/23289252-2399650
Fitzpatrick, S. J., & River, J. (2018). Beyond the medical model: Future directions for suicide intervention services. *International Journal of Health Services, 48*(1), 189–203.
Foucault, M., & Burchell, G. M. (2008). *The Birth of Biopolitics: Lectures at the Collège de France, 1978–1979*. Palgrave Macmillan UK. https://doi.org/10.1057/9780230594180
GATE. (2012). It's time for reform: Trans* health issues in the International Classification of Diseases. https://transactivists.org/94/
Gorton, N., & Grubb, H. M. (2014). General, Sexual, and Reproductive Health. In L. Erickson-Schroth (Ed.), *Trans bodies, trans selves: A resource for the transgender community* (pp. 215–239). Oxford University Press.
Hale, C. J. (1997, February 5). Suggested rules for non-transsexuals writing about transsexuals, transsexuality, transsexualism, or trans. *Sandy Stone*. https://sandystone.com/hale.rules.html
Hanssmann, C. (2016). Passing torches? Feminist inquiries and trans-health politics and practices. *Transgender Studies Quarterly, 3*(1–2), 120–136. https://doi.org/10.1215/23289252-3334283
Hill, A. O., Bourne, A., McNair, R., Carman, M., & Lyons, A. (2020). *Private Lives 3: A national survey of the health and wellbeing of LGBTIQ people in*

Australia. Australian Research Centre in Sex, Health & Society (ARCSHS), La Trobe University. https://www.latrobe.edu.au/arcshs/work/private-lives-3
Hunt, L. (2007). *Inventing human rights: A history*. WW Norton & Company.
Joffre, T. (2021, August 3). Transgender community to get more accessible healthcare—Health Ministry. *The Jerusalem Post (Online)*. https://www.jpost.com/health-science/transgender-community-to-get-more-accessible-healthcare-health-ministry-675716
Johnson, A. H. (2015). Normative accountability: How the medical model influences transgender identities and experiences. *Sociology Compass, 9*(9), 803–813. https://doi.org/10.1111/soc4.12297
Kara, S. (2017). *Gender is not an illness. How pathologizing trans people violates international human rights law*. GATE (Global Action for Trans Equality). https://transactivists.org/wp-content/uploads/2017/12/Gender-is-not-an-illness-GATE-.pdf
Lane, R. (2018). We are here to help. *TSQ: Transgender Studies Quarterly, 5*(2), 207–227. https://doi.org/10.1215/23289252-4348648
Maggic, M. (2017). Housewives making drugs. https://www.media.mit.edu/projects/housewives-making-drugs/overview/
Malatino, H. (2020). *Trans care*. University of Minnesota Press.
Moreton-Robinson, A. (2015). *The white possessive: Property, power, and Indigenous sovereignty*. University of Minnesota Press.
Namaste, V. (2000). *Invisible lives: The erasure of transsexual and transgendered people*. University of Chicago Press.
Otto, D. (2016). Gender and sexual diversity: A question of humanity? *Melbourne Journal of International Law, 17*(2), 477–488. https://doi.org/10.5040/9780755609147.ch002
Oudshoorn, N. (2003). *Beyond the natural body: An archaeology of sex hormones*. Routledge.
Park, A. (2019). Yogyakarta Plus 10: A demand for recognition of SOGIESC. *North Carolina Journal of International Law, 44*(2), 223–272.
Plemons, E. (2018). A capable surgeon and a willing electrologist: Challenges to the expansion of transgender surgical care in the United States. *Medical Anthropology Quarterly*. https://doi.org/10.1111/maq.12484
Puar, J. K. (2017). *The Right to Maim: Debility, Capacity, Disability*. Duke University Press.
Rosenberg, S., Duck-Chong, E., & Cook, T. (2021). Gender affirming surgery in Australia: An evidence brief. https://www.transhub.org.au/s/Gender-Affirming-Surgery_Evidence-Brief.pdf

Rubin, H. (2006). The logic of treatment. In S. Stryker & S. Whittle (Eds.), *The transgender studies reader* (pp. 482–498). Routledge. https://doi.org/10.4324/9780203955055

Ryan, H. (2022). *The women's house of detention: A queer history of a forgotten prison*. Hachette UK.

Schwend, A. S. (2020). Trans health care from a depathologization and human rights perspective. *Public Health Reviews, 41*(1), 1–17. https://doi.org/10.1186/s40985-020-0118-y

Schwend, A. S., Espineira, K., & Walters, P. C. (2014). Depathologization. *Transgender Studies Quarterly, 1*(1–2), 73–77. https://doi.org/10.1215/23289252-2399650

Serano, J. (2016). *Whipping girl: A transsexual woman on sexism and the scapegoating of femininity*. Hachette UK.

Siegel, R. B. (1997). Why equal protection no longer protects: The evolving forms of status-enforcing state action. *Stanford Law Review*, 1111–1148. https://doi.org/10.2307/1229249

Siotos, C., Neira, P. M., Lau, B. D., Stone, J. P., Page, J., Rosson, G. D., & Coon, D. (2018). Origins of gender affirmation surgery: The history of the first gender identity clinic in the United States at Johns Hopkins. *Annals of Plastic Surgery, 83*(2), 132–136. https://doi.org/10.1097/SAP.0000000000001684

Smith, L. T. (2021). *Decolonizing methodologies: Research and Indigenous peoples* (3rd ed.). Zed Books.

Spade, D. (2013). Intersectional resistance and law reform. *Signs: Journal of Women in Culture and Society, 38*(4), 1031–1055. https://doi.org/10.1086/669574

Spiro, A. (2022). *Experts hit back at claim transgender voice therapies aren't life-saving*. The Times of Israel. https://www.timesofisrael.com/experts-hit-back-at-claim-transgender-voice-therapies-arent-lifesaving/

Stanley, E. A., & Smith, N. (2016). *Captive genders: Trans embodiment and the prison industrial complex* (2nd ed.). AK Press.

Stone, S. (1991). The empire strikes back: A posttranssexual manifesto. In J. Epstein & K. Straub (Eds.), *In body guards: The cultural politics of gender ambiguity* (pp. 280–304). Routledge.

Stop Trans Pathologization (STP). (2010). *Best practices guide to trans health care in the national health system*. http://www.stp2012.info/STP_guide_health.pdf

Stryker, S. (1994). My words to Victor Frankenstein above the Village of Chamounix: Performing transgender rage. *GLQ: A Journal of Lesbian and Gay Studies, 1*(3), 237–254. https://doi.org/10.1215/10642684-1-3-237

Stryker, S. (2017). *Transgender history: The roots of today's revolution* (2nd ed.). Seal Press.
Tamboukou, M. (1999). Writing genealogies: An exploration of Foucault's strategies for doing research. *Discourse: Studies in the Cultural Politics of Education, 20*(2), 201–217. https://doi.org/10.1080/0159630990200202
Taylor, K. (2019). Pornography addiction: The fabrication of a transient sexual disease. *History of the Human Sciences, 32*(5), 56–83. https://doi.org/10.1177/0952695119854624
Trans Safety Network. (2024). TSN statement on Cass Review Final Report. https://transsafety.network/posts/tsn-statement-on-cass-final-report/
Tseris, E. (2019). Accepting my illness? Problematising the claims of mental health anti-stigma efforts. In D. Baines, B. Bennett, S. Goodwin, & M. Rawsthorne (Eds.), *Working across difference: Social work, social policy and social justice* (pp. 155–169). Red Globe Press.
Velocci, B. (2021). Standards of Care: Uncertainty and Risk in Harry Benjamin's Transsexual Classifications. TSQ: *Transgender Studies Quarterly, 8*(4), 462–480. https://doi.org/10.1215/23289252-9311060
Waites, M. (2009). Critique of 'sexual orientation' and 'gender identity' in human rights discourse: global queer politics beyond the Yogyakarta Principles. *Contemporary Politics, 15*(1), 137–156. https://doi.org/10.1080/13569770802709604
Weheliye, A. G. (2014). *Habeas viscus: Racializing assemblages, biopolitics, and black feminist theories of the human*. Duke University Press.
Winters, K. (2022). Contradictions and compromises of principle in the SOC8. https://transpolicyreform.files.wordpress.com/2023/02/20221 0soc8principlesj.1.pdf
World Health Organisation. (2020). *Classifications*. https://www.who.int/classifications/icd/en/
Worley, J. (2011). 'Street power' and the claiming of public space: San Francisco's 'Vanguard' and Pre-Stonewall queer radicalism. In N. Smith & E. A. Stanley (Eds.), *Captive genders: Trans embodiment and the prison industrial complex* (2nd ed., pp. 41–56). AK Press.
Yogyakarta Principles. (2007). *Yogyakarta principles: Principles on the application of international human rights law in relation to sexual orientation and gender identity*. http://data.unaids.org/pub/manual/2007/070517_yogyakarta_principles_en.pdf

Yogyakarta Principles. (2017). *Yogyakarta Principles Plus 10: Principles on the application of international human rights law in relation to sexual orientation, gender identity, gender expression and sex characteristics to complement the Yogyakarta Principles.* http://www.yogyakartaprinciples.org/principles-en/yp10/

11

Social Media Contagion, Digital Self-Diagnosis and Youth Mental Health

Natalie Ann Hendry

Introduction

Popular critics, moral entrepreneurs and psy-professionals continue to champion ever-evolving concerns about social media or digital technology provoking and spreading psychopathology in young people in Australia and elsewhere. Such popular conversations in news media and health research routinely ignore the relatively significant—and complicated—history of youth participation related to mental health on the internet, social networking and other digital media contexts and cultures. Obscuring this history overshadows how young people have long used media technologies to make sense of their experiences of mental health to build collective yet amorphous expertise that engages with and, at times, challenges, formalised psychiatric and psychological knowledges.

N. A. Hendry (✉)
Faculty of Education, University of Melbourne, Melbourne, VIC, Australia
e-mail: natalie.hendry@unimelb.edu.au

N. A. Hendry, E. Karageorgos (eds.), *Critical Mental Health in Australia and Aotearoa New Zealand*, The Politics of Mental Health and Illness,
https://doi.org/10.1007/978-3-031-82861-4_11

Instead, what does continue, again and again, to garner professional and institutional attention is the presumed potential that participating on and scrolling through social media platforms—especially when pseudonymous, unsurveilled and 'uncontained'—risks social contagion: the spontaneous spread of ideas or feelings through collective engagement online (Gansner et al., 2024; Haltigan et al., 2023), akin to fears of contagion on the 'therapeutic milieu' of a physical psychiatric ward. In such accounts, young people are positioned as highly suggestible to self-diagnosis and psychopathologisation via scrolling through social media content or interacting via digital technologies (Haltigan et al., 2023).

Teaching and working with young people in Australia for two decades and researching youth digital cultures and mental health for the last ten years leaves me hesitant when social media or digital technologies are framed as 'contagious' or when someone suggests that social media is uncontrollably spreading a particular idea or action about bodies, feelings or attitudes. I want to pause on accounts claiming that, for instance, young people *only* identify with attention-deficit/hyperactivity disorder (ADHD) or label themselves as having an anxiety disorder because of TikTok videos, Instagram Stories or other digital media. Digital media 'effects' are framed as contagious, where content and media practices become seen as vectors of infectious disease. Yet, these approaches ignore media studies and other cultural and social scholarship that offer more complex, less deterministic, theories of digital communication.

In the context of mental health, the psy-complex (Rose, 1985) and therapeutic cultures (Rimke, 2020; Salmenniemi et al., 2019; Wright, 2011), more attention is needed towards how we imagine, conceptualise and theorise what it means to engage and participate through digital media and cultures. I argue that it is unhelpful to make simple claims about digital 'exposure' or digital activities perpetuating self-diagnosis or mental illness. This is not to say that content or digital practices can never be disruptive, disturbing or prompt thoughts, feelings or actions.[1] But we cannot universally assume how people, especially young people, participate digitally and

[1] I also note the substantial humanities and social research scholarship engaging with affect theory and related approaches to understand affect and contagion through digital media (e.g., Blackman, 2007; Knudsen & Stage, 2012). Here, I am instead interested in the political and social deployment of contagion as a means of explaining media effects of digital technologies and media.

how this shapes how they think, feel or enact ideas or beliefs about mental health. Concepts or metaphors from psychology and psychiatry cannot be simply translated into a digital context, especially those that are based on deficit assumptions about what is 'normal' or 'rational' behaviour. If we are to develop a critical scholarship of digital media and mental health, one that is wary of psychiatric or psychological concepts and how they are deployed in popular vernacular, we might better embrace more nuanced theories and concepts of digital communication.

This chapter offers a small contribution towards such aims. It reviews how we theorise media effects in the context of 'social contagion' where contagion is an all-too-simple way of understanding how people may step outside the normal boundaries of 'sane' and 'mad'. Guided by media studies, sociology and cultural studies approaches, I hope to make more space for ambivalent or contradictory digital experiences that take into account broader social processes. The chapter begins by sharing some examples of how the concept of 'contagion' both explicitly and implicitly underpins assumptions about the relationship between social media and mental health. Given that much of this attention is directed towards young people, I ground my thinking in relation to youth mental health in the Global North. I reflect on how social contagion has been conceptualised and deployed in psychological and psychiatric research and how this has been translated as digital or social media contagion in the context of digital media and technology. The chapter outlines how other media theories provide more complex and nuanced perspectives, then offers some concluding thoughts about how this discussion might support critical mental health scholars to more closely interrogate how therapeutic cultures are sustained through digital media and technology.

Contagious (Self-)Diagnosis and 'Social Media Contagion'

Can we simply transplant psychiatric concepts into the digital world? Should we? In 2023, US developmental psychologist John D. Haltigan and colleagues published a short conceptual article in the journal,

Comprehensive Psychiatry. Writing that social media is akin to "an incubator" that induces people to self-diagnose psychiatric disorders, Haltigan et al. (2023, p. 1) explain that young people, particularly young women, are increasingly identifying with conditions such as depression and anxiety, autism, gender dysphoria, tics and Tourette's Syndrome, and eating disorders with no psychiatric or biological basis. They propose that this

> ostensibly heterogeneous sociogenic illness behavior, loosely bounded by classical notions of mental health diagnoses, suggests the possibility that the increasingly algorithmic and audiovisually immersive social media environment is a scopic medium in which various 'neurodivergent' or sick role identities or personas can be claimed at will, at any given moment—with no antecedent biological basis or tether to empirical reality—with positive social and emotional reinforcement and resonance from the associated online community (e.g., via the use of hashtags; user-to-user sharing and amplification of content). (Haltigan et al., 2023, p. 3)

Haltigan et al. (2023, p. 1) argue that there is "an urgent need" to understand contagion on social media and other platforms, given that "algorithmic social media platforms may serve as a vehicle of transmission for social contagion of self-diagnosed mental illness conditions" and that platforms like TikTok are a "spread vector" for false self-diagnoses. They suggest a *possibility* that social contagion influences identity confusion (this refers to transgender identity) and mental ill-health and cite references and links, for example, from editorial articles, studies that involved anti-trans websites to recruit participants, and (arguably) conservative commentators.[2] This perhaps points to the paper offering a political or cultural argument that exceeds the boundaries of what we might typically know as psychiatric scholarship.

[2] These citations include US physician Littman's (2018) article proposing the existence of "rapid-onset gender dysphoria" and the subsequent correction (2019) stating that the condition was not clinically validated (Trans Data Library, 2024); an article by Kathryn Dee, a media journalist who frequently publishes in conservative outlets; and a tweet by Wesley Yang who is perceived to subscribe to "anti-identity politics" (Lin, 2022). Some of these references relate to the trans history of the platform, Tumblr (see Haimson et al., 2019; Tiidenberg et al., 2021, for a broader discussion of gender diversity and Tumblr).

What stands out here about the relationships between social media and mental health is that digital communication related to (self-)diagnosis is viewed as contagious or mimetic—that it inherently 'grabs' you. Contagion conveys assumptions about the influence of digital media and its intensive power. This framing implies a normative judgement, that individuals, especially young women, who are susceptible to this contagious content are immature and guided by *too much* peer influence and they note that there are likely psychological or developmental reasons for why young women are more susceptible to contagious behaviour. Haltigan et al. (2023) also identify the "diathesis-stress model"—that conditions are caused by an inherent disposition alongside experiences of stress that trigger onset—as one potential explanation for why some people may be more susceptible and subsequently at risk of digitally influenced sociogenic conditions. However, this model has been challenged (Simons, 2022) and, more importantly, it ignores the complexity of how 'stress' is understood and experienced socially. As Brossard and Chandler (2022, pp. 62, original emphasis) offer, given "the diversity of biological phenomena attached to stress [...this] suggests that the shared diversity of human experience cannot be tackled *only* by differentiating several types of stress, nor by the isolation of 'stressors', 'buffers' and 'coping mechanisms'". In some ways, contagion acts as faulty or limited metaphor that can only explain *some* of these phenomena; social media as a powerful 'spread vector' enables anyone at any given moment to claim a mental diagnosis 'persona' but, at the same time, this likely happens to vulnerable young women and those with poor emotion regulation.

This concept of social media contagion has circulated internationally, particularly in response to concerns about youth mental health (e.g., Weigle, 2023), and popular commentary has also emerged in Australia. Much of the recent news media attention has focused on the social media platform TikTok (a short-video platform) and the increase in mental health-related content shared on the platform through the COVID-19 pandemic.[3] In 2022, popular Australian women's lifestyle website,

[3] The growth in TikTok traffic from Australia during the COVID-19 was also related to TikTok's history (Zeng et al., 2021). Chinese company ByteDance developed the short-video platform Douyin in 2016 and released TikTok as an international version of Douyin in 2017. ByteDance acquired the Chinese and US marketed Gen Z platform Musical.ly in 2017 and consolidated

Mamamia, reported on a "worrying" trend that young people were watching TikTok videos about mental health and seemingly misdiagnosing borderline personality disorder, ADHD and bipolar disorder among other disorders (Docherty, 2022). The article quotes psychologists speaking about the "dangers" of "less-than-credible" resources and contrasts this to a positive story about self-diagnosis that involved a mother using social media to make sense of her child's eventual ADHD diagnosis. It then positions TikTok as a source of sociogenic illness as it might prompt a false diagnosis. Like other popular news social media and mental health articles or posts, it ends with pleas for people to seek professional, expert, psychological or psychiatric advice. In the following year, the Australian Psychological Society (APS) reposted an article from the fashion magazine, *Harper's Bazaar* (2023), that asked if "social media has made therapy sexy" and quoted the APS president Dr Catriona Davis-McCabe saying that social media therapy or self-diagnosis "is no match for seeing a professional and getting a proper diagnosis". Other perspectives note that more and more Australians are self-diagnosing mental illness or disorders—again, seemingly due to TikTok and other short-form social media content and first-person accounts of mental ill-health—but they portray this as a problem stemming from limited and expensive access to formal psychiatric diagnosis and treatment and (perhaps in a more critical account) that pathologising human experience is increasingly common (Denny, 2023). These perspectives ignore that professional help itself is a social process (Brossard & Chandler, 2022). Critique focuses on containing expertise to psy-professionals and surveilling and controlling discourse around diagnosis from often pseudonymous and uncontained (or uncontainable) sources.

British psychologists Foulkes and Andrews (2023) suggest a bidirectional relationship between mental health awareness activities, like the above media articles and other public campaigns and advertising, and increases in mental health problems. They frame this two-way relationship as the "prevalence inflation hypothesis" (p. 1) that consists of two mechanisms. First, awareness campaigns and activities contribute to

---

TikTok and Musical.ly in 2018. Following attention in the youth market in its early years, TikTok gained popularity during COVID-19 lockdowns, which diversified the platforms' audience.

increases in mental health problems due to "improved recognition"—what awareness aims to do, to encourage people to seek help and decrease stigma related to mental ill-health—and second, the same campaigns encourage "overinterpretation", where people misinterpret their everyday experiences and understand their lives through misapplied psychological or psychiatric concepts or labels (Foulkes & Andrews, 2023, p. 1). This, they explain, might be due to a "self-fulfilling prophecy" or that mental health problems are romanticised and influenced by social media via "cultural commentators" (p. 2). Extending this logic, Foulkes and Andrews (2023) cite Hacking's (1996) looping theory, where classifications made by clinicians to describe behaviour become self-reinforcing as people change their behaviour to fit a classification's narrative, creating a dynamic looping effect between the categories and what is being categorised (for a useful discussion see Brossard & Chandler, 2022). They propose that similar processes are emerging—perhaps based on media exposure rather than explicitly noting contagion models—where people self-diagnose, as "individuals might more readily label themselves" (p. 2), which may be beneficial (e.g., prompt treatment or self-understanding) but may also encourage overinterpretation. Arguably, this perspective aligns well with critical mental health scholars' concerns about the psychologisation of everyday life and how psychiatric concepts have come to shape social practices and knowledges.

Yet in these accounts, whether the metaphor of contagion is explicit or implicit, social media or digital contagion and 'exposure' does little to bring more precision or nuance to discussions about how digital communication or (self-)diagnosis function or are experienced. Contagion here provides a model of sociality; that being in the same place, at the same time, is enough for peer influence to travel: for affects, moods, behaviours, ideas and beliefs to 'jump' from one social body to another. The relationships between social media platforms and mental health are variously framed as automatic, where appropriating experiences and thoughts as psychopathological happens too quickly and for invalid reasons. This erases diversity (e.g., all young people are subject to the contagion, more so those susceptible to it) and suggests that media engagement is coercive or limited only to that contagious content. Each of these examples allude to a parasitic framing of social media that sidesteps the cultural economies of both platform surveillance and mental health industries and discourses; the messiness and complexity of digital cultures are lost.

Contagion here denotes excess—an excessive feeling, excessive identification—that brings with it an assumption about inauthenticity: of being too much and needing intervention. These implied models of digital communication typically rely on long-standing, albeit long-outdated, models of communication. One such model, the 'hypodermic model' of communication, suggests there is a linear relationship between what a sender intends a message to say, and how the receiver interprets this (Thibault, 2016). Elsewhere with Kath Albury (Albury & Hendry, 2023), I have discussed how sexual health educators and clinical services narrowly interpret digital sexual health content as transmission or information seeking. We instead propose other theoretical approaches towards digital *cultures* of sexual health. In this chapter, I focus on contagion as another example of the inescapable 'viral' hypodermic needle model that potentially conflates all media communication as merely 'exposure'.

Social Contagion, Infectious Disease Metaphors and Disordered Crowds

As a group-oriented, seemingly clustered experience (Cheng et al., 2014), contagion is employed as a conceptual and material mechanism in clinical psychology and psychiatric research to explain how ideas, beliefs, emotional states or behaviours spread consciously and/or unconsciously between people and groups (Bastiampillai et al., 2013; Kindermann & Skinner, 2019). The concepts of social contagion or emotional contagion, for example, have helped researchers conceptualise and explain how mental health experiences like depression or happiness can spread between clusters of people or to "three degrees of influence" (Christakis & Fowler, 2013, p. 562) even though contagion is not simple to map or measure both in the moment or longitudinally (Bastiampillai et al., 2013). Contagion effects are understood, even within this literature, as not necessarily inevitable, predictable or straightforward, even for people and groups in close temporal proximity to each other (e.g., Eisenberg et al., 2013). In part, contagion theory is used to explain how psychogenic illnesses and disorders spread, where physical and/or psychological symptoms are thought to be transmitted between people without any biological cause (Warren & Power, 2015).

Theories and concepts about social and emotional contagion appear in youth research about suicide (e.g., Martínez et al., 2023), eating disorder behaviours (where attitudes aligned to so-called disordered eating are attributed to peer cliques and networks; e.g., Allison et al., 2014), self-injury or self-harm (e.g., Jarvi et al., 2013), and violence and mass school shootings (unsurprisingly this scholarship is focused in the United States; e.g., Bond & Bushman, 2017; Fox et al., 2021). The conditions for contagion effects for young people are typically framed in relation to how they may be more likely than adults to be (overly) oriented towards their friendship circles and networks, where gaining social status may be enhanced by adopting the poor social norms or behaviours of peers (Bastiampillai et al., 2013). This literature also relies on developmental theories of adolescence (in contrast to other approaches, see Wyn, 2009), whereby young people are understood as easily susceptible to risky ideas and inherently illogical, akin to earlier assumptions about the "storm and stress" of "adolescence" by psychologist Hall (1904) in the early twentieth century. Even though this concept has been challenged within and outside of psychology and psychiatry (e.g., Buchanan et al., 2023), it persists through neuroscientific claims of the impulsive and risky adolescent brain (Bessant, 2008; Kelly, 2012).

Given the flexibility of how contagion has been used in research and practice, Kindermann and Skinner (2019) argue that even within psychology the concept of contagion has become too imprecise and, in response, offer a more precise conceptualisation. Distinguishing contagion from other social processes of influence or social exchange, they aim to better align the concept in relation to "a specific mechanism of disease transmission" (p. 741). The essential elements of contagion, in their reading, include

> (a) a one-way transmission of an attribute, from someone who has it (Person A) to someone who does not (Person B), that (b) does not result in

> Person A losing the attribute or (c) in Person A receiving anything from Person B in return, and (d) does not involve the transmission of the absence of the attribute as well. (p. 743)

They propose that this contrasts contagion from other forms of social influence such as transactions, transfer, exchange or peer pressure. The temporal elements of this framework for contagion suggest that over time, more and more people are required to be "infected" and "Infection processes cannot cause problematic characteristics to decrease" (Kindermann & Skinner, 2019, p. 746). This approach understands contagion to grow exponentially, at least for a particular period of time. What is unclear here, though, is how contagion slows or changes over time.

In the context of psychiatric treatment, the physical site or space for contagion to occur is typically understood as the site of the 'therapeutic milieu'. The therapeutic milieu refers to physical, cultural and therapeutic spaces that are oriented towards psychiatric observation, treatment and service delivery. These are spaces in the context of hospitalisation, the "social climate of the unit" or a "ward atmosphere, a product of interactions between the psychological, physical, social, political, and spiritual components" (Canvin et al., 2024, p. 215). Health research has framed the therapeutic milieu as a space of safety as it purports to provide physical containment at a particular physical site, with predictable structure and activities, and privileges interactions with staff and patients characterised by care and validation (Canvin et al., 2024).

For young people in particular, staff working on the ward typically direct their psychiatric attention towards young people's social interactions within the milieu, where socialising is monitored, evaluated (e.g., 'good' peer interactions are noted as part of progress and recovery), constrained and, at times, actively restricted. Staff fears of contagious aggression or other asocial activity (e.g., convincing someone else not to take their medicine) are justified by their beliefs and experiences that aggression, violence or self-harm beget the same behaviours from others and occur in clusters (Beck et al., 2018). Attention towards the potential for social contagion to occur on youth hospital ward milieux persists as a practised knowledge, even as formal research is unclear about how the therapeutic milieu contributes to social contagion (Canvin et al., 2024).[4] Any fears about contagion, and subsequent limits on peer interaction that staff want to avoid, also ignore that young people routinely stress that being with other peers 'who get it' is the central most useful and supportive element of their inpatient admissions (Moses, 2011).

We also might position this 'common sense' or moral logic of the physical, social and emotional safety of the therapeutic milieu as dubious. As critical histories of survivor-consumer experiences, from Mad Studies scholars, and other contributors to this book (see Chap. 6, Sinclair, Chap. 3, Tseris), ward atmospheres are regularly experienced as violent, dehumanising or unhelpful by consumers both during and long after their admissions (Daya, 2022; Leggatt & Jeffs, 2021; McMahon, 2023; Tseris et al., 2024). These critiques also find further support in social research highlighting the paradox that a predictable, contained milieu may not be conditional to emotional and social safety even if this is desired by staff and clinicians. As Canvin et al. (2024, p. 222) highlight, quoting a ward manager, "the more restrictions we put on people, the more chance we've got of violence". Any disruptions or safety incidents may rather be examples of events emerging from synchronicity or the "same spatio-temporal context" (p. 226) rather than contagion per se.[5]

However, the concept and metaphor of contagion has a history much longer than hospital ward milieux or youth psychiatry. The roots of social contagion theory emerged in the late nineteenth century as early European psychologists increasingly paid attention towards crowds, imitation and 'moral contagion' (Mitchell, 2012). While a full overview of these developments is beyond the scope of this chapter, I briefly note some of this work as it demonstrates how claims of contagion were more often claims

[4] Here, I also draw on my own experiences working as a teacher in a youth psychiatric hospital as well as conversations with colleagues still working, often ambivalently, in the sector. At the same time, I recognise that different inpatient contexts will have different norms and values related to social contagion and peer interaction.

[5] Moses (2011, p. 134) similarly find that restrictions do little to support care for young people who are hospitalised as these young consumers "expressed varying degrees of disenchantment and perplexity with rules and practices that not only made little sense but also were perceived as counter-therapeutic, particularly excessive 'alone time', a rigid schedule, restrictions on interpersonal interactions with peers, and specific stipulations of the point-level system [where different levels of risk are allocated points that align with different levels of freedom]".

of assumed disorder and deviance, just as we have seen in relation to the youth hospital milieu, rather than analyses of how ideas or practices were communicated or transmitted.

In the late nineteenth century, French psychologist Le Bon (1896) offered an enduring concept of the crowd and crowd behaviour. For Le Bon, people in a crowd demonstrate deindividuation or lose their individual, logical self in favour of collective, often disordered behaviour or a group mentality, whereby

> the disappearance of the conscious personality, the predominance of the unconscious personality, the turning by means of suggestion and contagion of feelings and ideas in an identical direction, the tendency to immediately transform the suggested ideas into acts; these, we see, are the principal characteristics of the individual forming part of a crowd. He is no longer himself, but has become an automaton who has ceased to be guided by his will. (p. 19)

In their sketch of crowd psychology and science before the mid-twentieth century (prior to more productive developments in intergroup and social identity theories), social psychologists Drury and Stott (2011, p. 278) stress that theories, like Le Bon's, were limited as

> the early 'crowd scientists' looked at 'the crowd' as if it were an isolated specimen on a Petri dish. By neglecting the social context of crowd violence, they were inevitably drawn to the conclusion that it emanated solely from sources within the crowd itself.

What is notable to Drury and Stott (2011) is that the social context of these theorists greatly influenced how they understood crowds and groups. Accounts of the crowd as irrational or deviant featured prominently in theory at times when crowds were positioned socially and politically as "a threat to the existing order" (p. 286). Crowd contagion theories conveyed moral discernment, perhaps more so than they conveyed a theory of transmission.

In a different example, we can see that imitation and contagion were understood as models of social activity, rather than social disorder. In the same period as Le Bon, French sociologist Gabriel Tarde offered a theory of contagion. Proposing that social proximity turns people into an (often) irrational, imitative crowd, Tarde (1903) was particularly interested in people's micro, interpersonal relationships and, importantly, the networks that enabled contagion and ideas, practices and beliefs to spread. As Mitchell (2012) notes, while Tarde acknowledged that modernity and urbanisation had brought with it changes to the earlier "passive obedience to ancestral orders, customs, and influence", this obedience remained but had been neutralised through people's "submission to the pressure, advice, and suggestions of contemporaries" (Tarde, 1903, p. 246). Tarde's approach took seriously that social influence was networked, imitative and directed by social power. For us, this calls for attention to the networks that structure the suggestible capacity of powerful 'contemporaries' for provoking imitation and contagion—whether these networks are the structure of the therapeutic milieu or networks of social media platforms, for example.

From Social Crowds to Digital Contagion, Networks and Interactivity

Returning to self-diagnosis, the concerns that Haltigan et al. (2023) raise towards the supposedly contagious capacity of media to foster self-diagnosis are not new nor recent. Translating contagion from the social to the digital recalls that the history of contagion has long pointed to media as a provocative source or infectious carrier of contagion. In 1774, Goethe's publication of the novel, *The Sorrows of Young Werther*, was perceived to prompt an increase in suicide by the novel's young readers who connected to the book's romantic, suicidal hero (Ferreday, 2010). This early phenomenon of 'suicide contagion' continues to be cited in medical and psychiatric articles about social media, young people and mental health (e.g., Gansner et al., 2024; Weigle & Shafi, 2024), and the role of media continues to be entangled in how suicide clusters in youth are studied (e.g., see Swedo et al., 2021, writing about a suicide cluster in the United States partly prompted young people posting about suicide on social media and the Netflix series, *13 Reasons Why*). Ferreday (2010) notes that media coverage of youth suicide engages a discourse where young people's digital media engagement is always-already excessive and pathological, thus ignoring more hopeful or even ambivalent readings of media and mental health.

While Haltigan and colleagues point to Tumblr as a precursor to TikTok-as-disorder-incubator, there is a longer history exploring the relationship between digital or internet-based media and mental health outcomes; this includes, for instance, studies on self-injury, eating disorders, suicide and depression on pre-TikTok sites and platforms such as LiveJournal, MySpace, Flickr, blogs and discussion forums (Gailey, 2009; Giles & Newbold, 2011; Seko, 2013; Zinoviev et al., 2016). This research suggests that the relationship between the digital and the psychiatric is complicated, and rarely can the 'impacts' of digital life easily be identified or assumed as universal as each site, platform, app or digital practice functions differently for different groups.

The Facebook emotional contagion experiment by Facebook engineers and psychologists at Cornell University is perhaps the most well-known recent example of digital contagion as a means to conceptualise influence *en masse* (Kramer et al., 2014). As Banner (2016) notes, while there were concerns raised about the study's lack of informed consent—Facebook users were unaware that positive posts in their newsfeed were being manipulated to determine if this would influence their emotions—there were also questions asked about the notion of 'emotional contagion' and whether it was an accurate or even useful concept for studying digital influence. In response, Facebook and other platforms attempt to gain legitimacy by establishing themselves as public health figures, developing suicide prevention initiatives that adopt "the rhetoric of contagion familiar to public health professionals" (Banner, 2016, p. 96). These initiatives reinforce the social power of the psy-disciplines as they inevitably involve digital surveillance or tracking by the platforms and public health and psychiatric organisations.

So far, we can see how contagion in psychology and psychiatry—that ideas, behaviours, feelings or attitudes can spread between people seemingly like a virus or disease—acts as both a concept and metaphor to explain social influence, both within and outside digital contexts. Even as the concept or phenomenon is challenged by empirical studies (e.g., Buchanan et al., 2023), it still persists and influences how practitioners and psy-researchers interpret media and social interaction. Thinking about other social and critical approaches to mental health, I see my argument reflecting social research related to diagnoses and labelling. Models of contagion and digital self-diagnosis as a contagious practice ignore much of the social processes of diagnosis and labelling. While this is a substantial body of research to summarise, one that is beyond this chapter, we can take up Brossard and Chandler's (2022, p. 89) excellent discussion of labelling theory in sociology as they remind us that "Labels do not solely qualify mental health disorders. They contribute to producing these disorders, generating the material and symbolic worlds that people—diagnosed or not—inhabit". Likewise, critical scholars are unlikely to disregard diagnosis as a social process, as outlined in Blaxter's (1978) paper about alcoholism where diagnosis is a category and process, and then later developed by Jutel (2009) as the sociology of diagnosis. In contrast, contagion as a model simplifies and erases the social processes that produce how we understand and interpret behaviour and knowledge through labels and diagnoses.

We can also imagine digital contagion as a kind of folk theory of how digital life operates. Young people in my research (Hendry, 2020) and teaching workshops exploring social media and mental health (Hendry, 2017) often attribute online harms and problems to stem from individuals' behaviours or beliefs rather than social or platform conditions (e.g., how social media platforms structure interaction in favour of commercial ends), and in turn, maintain that individuals should be responsible for fixing issues. Young et al. (2023) find that the folk theories—the symbolic resources, 'common sense' and shared moral vernacular about how people explain and make sense of their worlds—shape how young adults interpret and make moral judgements and predictions about their digital lives and other people's behaviours. Again, we find contagion as a concept or metaphor that not only makes claims about communication or transmission, but one that also produces normative distinctions about people, groups, values, ideas and behaviours.

What imaginations and forms of sociality are invoked, then, when we rely on contagion, or even simple exposure or viral spread, to understand

and explain the relationships between mental health and digital media? In *Contagious metaphor*, Australia media scholar Peta Mitchell (2012) explores how contagion has broad social and cultural utility. Extending finance scholar Robert W. Kolb's (2011) analysis of contagion in economics (where contagion is a threat to rational financial models), Mitchell (2012, p. 134) posits that contagion implies disorder, disruption and excess, beyond what is expected or normal; "contagion – like metaphor – is seen both to assist and to thwart attempts to explain the seemingly inexplicable". This echoes earlier points in this chapter, where contagion is seen as a normative concept of communication or transmission.

In Mitchell's (2012) analysis, the perceived phenomenon and the metaphor of contagion bring with them the concept of the network; networks are a condition of contagious transmission. While contagion relies on the network—she explains this via the 2011 virus-turned-pandemic film, *Contagion*—so too does the network rely on contagion. We can apply this to our earlier discussion of social media as an incubator for sociogenic illness. Assumptions about digital contagion imply a type of digital network that is all-encompassing and seemingly existing in a vacuum where the only network that matters is the network of mental health 'contagion' content circulating via the TikTok algorithm or Instagram feed.

If we embrace Mitchell's reasoning, attending to contagion (or communication or influence and so on) invites us to consider the networks that organise communication. In media studies and related disciplines, attention to networks can often focus on how digital media is structured and subsequently how people interact with the features, affordances or capacities of technology. However, as American information studies scholar Jennifer Stromer-Galley (2004) highlights, "interactivity" in digital research can refer *both* to how people interact with each other through digital media *and* also how they interact with devices, platforms and networks. The first she understands as "interactivity-as-process [which] entails a research focus on human interaction" (p. 392); the second considers "interactivity-as-product" and "user interactions with technology" (p. 392). As Stromer-Galley (2004, p. 393) notes, the two concepts of interactivity are often conflated or built into one theory "at the expense of fully recognizing (and appreciating) the differences between human and media interaction" and this "occludes an ability to see that product interacts with process; that is, it is possible that the degree or features of medium interactivity might affect outcome variables of human interaction". If we apply these insights to simple depictions of social media contagion, we can see how *both* interaction with other people *and* interactions with algorithms, dashboards, feeds or platforms are important considerations for teasing apart assumptions about social media and mental health. For instance, researching how young people talk with each other about TikTok's 'personal' algorithm (Avella, 2023) or how they interpret TikTok for mental health support in contrast to less safe spaces (Byron, 2024), is just as critical as studying the TikTok algorithms that organise the order of videos shown to users. Digital interactivity may produce self-diagnosis as a *product* or contagion, but contagion is insufficient to explain interactivity as a *process* or how digital communication functions.

Central to my argument are media and communication theories that do not presume a stimulus-behaviour or sender/message/receiver model of communication. One seminal example of this is cultural theorist and sociologist Stuart Hall's encoding/decoding model that adopts a semiotic understanding of communication, where communication is not linear or necessarily predictable. This is distinct from psychological media effects, theories and concepts (see for example Valkenburg et al., 2016). Hall (1980) outlines four "relatively autonomous" (p. 126) stages of communication: "production, circulation, distribution/consumption, reproduction" (p. 117) where each stage influences and limits but does not determine the next. This means that social media content related to ADHD on TikTok, for example, may be 'read' differently when it is produced by a content creator and then uploaded, to when it is organised by TikTok's algorithms and circulated, or later when it is distributed as a user scrolls through the platform. In Hall's (1980) model, communication, content or messages are read within three hypothetical positions: the dominant-hegemonic (the communicated message is decoded as the sender encoded and intended it to be), the negotiated (decoding is a mix of both dominant and oppositional readings of the encoded message) and the oppositional (the message is decoded in a way that resists the sender's intended message). These positions are not distinct categories, however, as media viewers and users can move between different positions in response to broader social and cultural contexts and discourses. This challenges research that collects examples of misinformation or overinterpretation about mental disorders on social media platforms as being *sufficient* evidence that social media prompts people to identify with misinformation. For critical scholars, this stresses to us that therapeutic cultures and psy-discourses may invite dominant readings but that this does not preclude negotiated or oppositional decoding.

American media studies scholar Adrienne Shaw (2017) usefully extends Hall's model in the context of interactive media technologies and highlights that digital technologies expect and shape particular user interaction based on their design and environments and the *affordances* of media and technologies. Affordances are interpretive possibilities for what an object or space can do; these can be actions that an object was designed for (e.g., a photography-sharing platform that helps people connect through sharing photos), but also hidden affordances (e.g., a platform that excludes self-injury images based on assumptions about 'good' mental health) or imagined affordances (e.g., the contagious capacity of the platform to challenge psy-expertise as users find ways around restrictions for posting self-injury images; see Lindgren & Johansson, 2023, for a discussion of this in relation to Instagram). Shaw (2017, p. 596) suggests that like encoding/decoding, the meanings of affordances are "negotiated in particular moments" (p. 596). Given the interactive capacity of digital media and the blurred boundaries between producers and consumers, "we have to rethink the political implications of audience activity somewhat differently. All activity is not resistive, of course, but neither is it complicit" (Shaw, 2017, p. 600). In Haltigan et al. (2023) and other articles espousing social media contagion, we can start to identify assumptions about TikTok or other platforms having a powerful deterministic influence on young people's ideas and behaviours. In contrast, what more productive media theories offer are narratives that more closely allow us to consider if, when and how digital cultures and the affordances of digital media influence the social process of self-diagnosis.

A Question of Psy-Expansionism?

In this chapter I have argued that the concept or metaphor of 'contagion' is insufficient for understanding how mental health communication and influence happens in a digital world. As a normative concept, contagion often implies moral judgement or 'disorder' that may be self-fulfilling rather than productive for theorising digital influence. In response, I have offered examples of media studies approaches as starting points for more nuanced and complex analysis of media relationships that take into account the broader social and technological contexts of *both* mental health and digital media.

Attention to how psychological and psychiatric researchers and clinicians embrace the concept of social media contagion also invites us to think more carefully about how therapeutic cultures circulate, move and transform through digital media cultures. This question of the growth of therapeutic cultures and psy-expansionism runs along the political lines of the left and the right, given the discursive flexibility of the psychologisation of everyday life. This psychologisation both seemingly accounts for 'snowflake' children being coaxed into fake disorder identities (perhaps akin to Haltigan et al.'s reading), while also accounting for the ever-growing capitalist commercialisation and responsibilisation of our 'inner selves' and collective lives that critical mental health scholars seek to analyse.

My thinking here has been influenced by Scottish sociologist Julie Brownlie's discussion of social research into therapy. Claiming that much of the "social has slipped away", Brownlie (2004, p. 516) proposes that interaction is often ignored in sociological research that starts and ends with Foucauldian perspectives that, at times, does not empirically engage with resistance or processes of interaction. Instead, we might want to interrogate interaction—for Brownlie, analysing interaction between young people in a therapeutic programme and therapists; for me, researching interaction between young people who engage with social media and their digital mental health cultures. We might want to explore theories of communication and interaction with more space for nuance than contagion models and ask if and how young people may not always self-diagnose or wholly embrace a therapeutic sensibility or orientation that adopts the ideas, knowledges, attitudes and practices that we attribute to self-help culture, therapeutic culture or the psychologisation of everyday life.

I invite critical mental health scholars to also hesitate when psychiatric or psychological language moves from clinical practice into popular vernacular and to carefully interrogate *how* psy-discourses and therapeutic cultures circulate—to engage in a more fine-grained analysis of digital media and sociality in relation to the psychologisation of everyday life. This chapter, as a provocation, is as much a challenge to my own research as it is to anyone else. Therefore, how might we better study the complexity of life under digital capitalism and the psychologisation of everyday life without adopting limiting and moralising psychiatric interpretations of digital cultures?

References

Albury, K., & Hendry, N. A. (2023). Information, influence, ritual, participation: Defining digital sexual health. *Journal of Sociology, 59*(3), 628–645. https://doi.org/10.1177/14407833221136579

Allison, S., Warin, M., & Bastiampillai, T. (2014). Anorexia nervosa and social contagion: Clinical implications. *Australian & New Zealand Journal of Psychiatry, 48*(2), 116–120. https://doi.org/10.1177/0004867413502092

Avella, H. (2023). "TikTok ≠ therapy": Mediating mental health and algorithmic mood disorders. *New Media & Society, 146144482211472*. https://doi.org/10.1177/14614448221147284

Banner, O. (2016). Immunizing the social network: Public health and the "troubled teenager" in digital media. In K. Nixon & L. Servitje (Eds.), *Endemic: Essays in contagion theory* (pp. 95–119). Palgrave Macmillan.

Bastiampillai, T., Allison, S., & Chan, S. (2013). Is depression contagious? The importance of social networks and the implications of contagion theory. *Australian & New Zealand Journal of Psychiatry, 47*(4), 299–303. https://doi.org/10.1177/0004867412471437

Beck, N. C., Tubbesing, T., Lewey, J. H., Ji, P., Menditto, A. A., & Robbins, S. B. (2018). Contagion of violence and self-harm behaviors on a psychiatric

ward. *The Journal of Forensic Psychiatry & Psychology, 29*(6), 989–1006. https://doi.org/10.1080/14789949.2018.1516230
Bessant, J. (2008). Hard wired for risk: Neurological science, 'the adolescent brain' and developmental theory. *Journal of Youth Studies, 11*(3), 347–360. https://doi.org/10.1080/13676260801948387
Blackman, L. (2007). Is happiness contagious? *New Formations, 63*, 15–32.
Blaxter, M. (1978). Diagnosis as category and process: The case of alcoholism. *Social Science & Medicine, 12*, 9–17.
Bond, R. M., & Bushman, B. J. (2017). The contagious spread of violence among US adolescents through social networks. *American Journal of Public Health, 107*(2), 288–294. https://doi.org/10.2105/AJPH.2016.303550
Brossard, B., & Chandler, A. (2022). *Explaining mental illness: Sociological perspectives*. Bristol University Press.
Brownlie, J. (2004). Tasting the witches' brew: Foucault and therapeutic practices. *Sociology, 38*(3), 515–532. https://doi.org/10.1177/0038038504043216
Buchanan, C. M., Romer, D., Wray-Lake, L., & Butler-Barnes, S. T. (2023). Editorial: Adolescent storm and stress: A 21st century evaluation. *Frontiers in Psychology, 14*. https://doi.org/10.3389/fpsyg.2023.1257641
Byron, P. (2024). TikTok's queer public culture of mental health support. *First Monday*. https://doi.org/10.5210/fm.v29i5.13258
Canvin, K., Brierley-Jones, L., Ramsey, L., Baker, J., & Berzins, K. (2024). Safety and contagion in acute psychiatric wards: How the milieu is implicated in the occurrence of clustered safety incidents. *Theory & Psychology, 34*(2), 214–232. https://doi.org/10.1177/09593543231225636
Cheng, Q., Li, H., Silenzio, V., & Caine, E. D. (2014). Suicide contagion: A systematic review of definitions and research utility. *PLoS ONE, 9*(9), e108724. https://doi.org/10.1371/journal.pone.0108724
Christakis, N. A., & Fowler, J. H. (2013). Social contagion theory: Examining dynamic social networks and human behavior. *Statistics in Medicine, 32*(4), 556–577. https://doi.org/10.1002/sim.5408
Daya, I. (2022). Russian dolls and epistemic crypts: A lived experience reflection on epistemic injustice and psychiatric confinement. *Incarceration, 3*(2), 26326663221103445. https://doi.org/10.1177/26326663221103445
Denny, B. (2023, November 9). Self-diagnosis is on the rise, but is TikTok really to blame? *The Sydney Morning Herald*. https://www.smh.com.au/lifestyle/health-and-wellness/self-diagnosis-is-on-the-rise-but-is-tiktok-really-to-blame-20231026-p5efdi.html

Docherty, E. (2022, April 29). TikTok used to self-diagnose personality disorders. *Mamamia*. https://www.mamamia.com.au/personality-disorders-tiktok/

Drury, J., & Stott, C. (2011). Contextualising the crowd in contemporary social science. *Contemporary Social Science, 6*(3), 275–288. https://doi.org/10.1080/21582041.2011.625626

Eisenberg, D., Golberstein, E., Whitlock, J. L., & Downs, M. F. (2013). Social contagion of mental health: Evidence from college roommates. *Health Economics, 22*(8), 965–986. https://doi.org/10.1002/hec.2873

Ferreday, D. (2010). Reading disorders: Online suicide and the death of hope. *Journal for Cultural Research, 14*(4), 409–426. https://doi.org/10.1080/14797581003765366

Foulkes, L., & Andrews, J. L. (2023). Are mental health awareness efforts contributing to the rise in reported mental health problems? A call to test the prevalence inflation hypothesis. *New Ideas in Psychology, 69*, 101010. https://doi.org/10.1016/j.newideapsych.2023.101010

Fox, J. A., Sanders, N. E., Fridel, E. E., Duwe, G., & Rocque, M. (2021). The contagion of mass shootings: The interdependence of large-scale massacres and mass media coverage. *Statistics and Public Policy, 8*(1), 53–66. https://doi.org/10.1080/2330443X.2021.1932645

Gailey, J. A. (2009). "Starving is the most fun a girl can have": The pro-ana subculture as edgework. *Critical Criminology, 17*(2), 93–108. https://doi.org/10.1007/s10612-009-9074-z

Gansner, M., Berson, C., & Javed, Z. (2024). Social media contagion of high-risk behaviors in youth. *Pediatric Clinics of North America, S0031395524001330*. https://doi.org/10.1016/j.pcl.2024.07.037

Giles, D. C., & Newbold, J. (2011). Self- and other-diagnosis in user-led mental health online communities. *Qualitative Health Research, 21*(3), 419–428. https://doi.org/10.1177/1049732310381388

Hacking, I. (1996). The looping effects of human kinds. In D. Sperber, D. Premack, & A. J. Premack (Eds.), *Causal cognition: A multidisciplinary debate* (pp. 351–394). Oxford University Press.

Haimson, O. L., Dame-Griff, A., Capello, E., & Richter, Z. (2019). Tumblr was a trans technology: The meaning, importance, history, and future of trans technologies. *Feminist Media Studies*, 1–17. https://doi.org/10.1080/14680777.2019.1678505

Hall, G. S. (1904). *Adolescence its psychology and its relations to physiology, anthropology, sociology sex, crime, religion and education* (Vol. 2). D. Appleton & Company.

Hall, S. (1980). Encoding/decoding. In S. Hall, D. Hobson, A. Lowe, & P. Willis (Eds.), *Culture, media, language* (pp. 117–127). Routledge.

Haltigan, J. D., Pringsheim, T. M., & Rajkumar, G. (2023). Social media as an incubator of personality and behavioral psychopathology: Symptom and disorder authenticity or psychosomatic social contagion? *Comprehensive Psychiatry, 121*, 152362. https://doi.org/10.1016/j.comppsych.2022.152362

Harper's Bazaar. (2023, September 14). *TherapyTok: Social media has made therapy sexy, but what do we make of that?* https://psychology.org.au/about-us/news-and-media/aps-in-the-media/2023/therapytok-social-media-has-made-therapy-sexy,-but

Hendry, N. A. (2017). Social media bodies: Revealing the entanglement of sexual wellbeing, mental health and social media in education. In L. Allen & M. L. Rasmussen (Eds.), *Palgrave handbook of sex education* (pp. 509–526). Palgrave Macmillan.

Hendry, N. A. (2020). Young women's mental illness and (in-)visible social media practices of control and emotional recognition. *Social Media + Society, 6*(4), 1–10. https://doi.org/10.1177/2056305120963832

Jarvi, S., Jackson, B., Swenson, L., & Crawford, H. (2013). The impact of social contagion on non-suicidal self-injury: A review of the literature. *Archives of Suicide Research, 17*(1), 1–19. https://doi.org/10.1080/13811118.2013.748404

Jutel, A. (2009). Sociology of diagnosis: A preliminary review. *Sociology of Health & Illness, 31*(2), 278–299.

Kelly, P. (2012). The brain in the jar: A critique of discourses of adolescent brain development. *Journal of Youth Studies, 15*(7), 944–959. https://doi.org/10.1080/13676261.2012.693596

Kindermann, T. A., & Skinner, E. A. (2019). Is psychology suffering from an epidemic of "contagion"? Moving from metaphors to theoretically derived concepts and methods in the study of social influences. *Theory & Psychology, 29*(6), 739–756. https://doi.org/10.1177/0959354319857154

Knudsen, B. T., & Stage, C. (2012). Contagious bodies. An investigation of affective and discursive strategies in contemporary online activism. *Emotion, Space and Society, 5*(3), 148–155. https://doi.org/10.1016/j.emospa.2011.08.004

Kolb, R. W. (2011). What is financial contagion. In R. W. Kolb (Ed.), *Financial contagion: The viral threat to the wealth of nations* (pp. 3–10). John Wiley.

Kramer, A. D. I., Guillory, J. E., & Hancock, J. T. (2014). Experimental evidence of massive-scale emotional contagion through social networks.

Proceedings of the National Academy of Sciences, 111(24), 8788–8790. https://doi.org/10.1073/pnas.1320040111

Le Bon, G. (1896). *The crowd: A study of the popular mind.* Batoche Books.

Leggatt, M., & Jeffs, S. (2021). *Out of the madhouse: From asylums to caring community?* Australian Scholarly Publishing.

Lin, A. (2022, March 27). The former essayist: One the decline of Wesley Yang. *Strange Matters.* https://strangematters.coop/wesley-yang-souls-of-yellow-folk-review/

Lindgren, S., & Johansson, A. (2023). Getting better? Hegemonic, negotiated and oppositional uses of Instagram for mental health support. *Journal of Communication Inquiry, 47*(3), 268–290. https://doi.org/10.1177/01968599211042581

Littman, L. (2018). Parent reports of adolescents and young adults perceived to show signs of a rapid onset of gender dysphoria. *PLOS ONE, 13*(8), e0202330. https://doi.org/10.1371/journal.pone.0202330

Littman, L. (2019). Correction: Parent reports of adolescents and young adults perceived to show signs of a rapid onset of gender dysphoria. *PLOS ONE, 14*(3), e0214157. https://doi.org/10.1371/journal.pone.0214157

Martínez, V., Jiménez-Molina, Á., & Gerber, M. M. (2023). Social contagion, violence, and suicide among adolescents. *Current Opinion in Psychiatry, 36*(3), 237. https://doi.org/10.1097/YCO.0000000000000858

McMahon, R. (2023). Stories of the Silenced Manifesto and Mad Studies: An experience of biomedicine and mental health. *International Mad Studies Journal, 1*(1), 1–17. https://doi.org/10.58544/imsj.v1i1.5250

Mitchell, P. (2012). *Contagious metaphor*. Bloomsbury.

Moses, T. (2011). Adolescents' perspectives about brief psychiatric hospitalization: What is helpful and what is not? *Psychiatric Quarterly, 82*(2), 121–137. https://doi.org/10.1007/s11126-010-9151-1

Rimke, H. (2020). Self-help, therapeutic industries, and neoliberalism. In D. Nehring, O. J. Madsen, E. Cabanas, C. Mills, & D. Kerrigan (Eds.), *The Routledge International Handbook of Global Therapeutic Cultures* (1st ed., pp. 37–50). Routledge.

Rose, N. (1985). *The psychological complex: Psychology, politics and society in England 1869–1939.* Routledge.

Salmenniemi, S., Nurmi, J., Perheentupa, I., & Bergroth, H. (Eds.). (2019). *Assembling therapeutics: Cultures, politics and materiality* (1st ed.). Routledge.

Seko, Y. (2013). Picturesque wounds: A multimodal analysis of self-injury photographs on Flickr. *Forum Qualitative Sozialforschung/Forum: Qualitative Social Research, 14*(2), 1–23. https://doi.org/10.17169/fqs-14.2.1935

Shaw, A. (2017). Encoding and decoding affordances: Stuart Hall and interactive media technologies. *Media, Culture & Society, 39*(4), 592–602. https://doi.org/10.1177/0163443717692741

Simons, P. (2022, August 8). Study contradicts diathesis-stress model of psychosis. *Mad In America*. https://www.madinamerica.com/2022/08/study-contradicts-diathesis-stress-model-psychosis/

Stromer-Galley, J. (2004). Interactivity-as-product and interactivity-as-process. *The Information Society, 20*(5), 391–394. https://doi.org/10.1080/01972240490508081

Swedo, E. A., Beauregard, J. L., Fijter, S. de, Werhan, L., Norris, K., Montgomery, M. P., Rose, E. B., David-Ferdon, C., Massetti, G. M., Hillis, S. D., & Sumner, S. A. (2021). Associations between social media and suicidal behaviors during a youth suicide cluster in Ohio. *Journal of Adolescent Health, 68*(2), 308–316. https://doi.org/10.1016/j.jadohealth.2020.05.049

Tarde, G. (1903). *The laws of imitation* (E. C. Parsons, Trans.). Henry Holt.

Thibault, G. (2016). Needles and bullets: Media theory, medicine, and propaganda, 1910–1940. In K. Nixon & L. Servitje (Eds.), *Endemic: Essays in contagion theory* (pp. 67–92). Palgrave Macmillan.

Tiidenberg, K., Hendry, N. A., & Abidin, C. (2021). *Tumblr*. Polity Press.

Trans Data Library. (2024, January 21). *Lisa Littman*. Trans Data Library. https://transdatalibrary.org/person/lisa-littman

Tseris, E., Franks, S., & Hart, E. B. (2024). *Psychiatric oppression in women's lives: Creative resistance and collective dissent*. Palgrave Macmillan.

Valkenburg, P. M., Peter, J., & Walther, J. B. (2016). Media effects: Theory and research. *Annual Review of Psychology, 67*(1), 315–338. https://doi.org/10.1146/annurev-psych-122414-033608

Warren, Z. J., & Power, S. A. (2015). It's contagious: Rethinking a metaphor dialogically. *Culture & Psychology, 21*(3), 359–379. https://doi.org/10.1177/1354067X15601190

Weigle, P. E. (2023, May 8). Psychoeducation or psychiatric contagion? Social media and self-diagnosis. *Psychiatric Times*. https://www.psychiatrictimes.com/view/psychoeducation-or-psychiatric-contagion-social-media-and-self-diagnosis

Weigle, P. E., & Shafi, R. M. A. (2024). Social media and youth mental health. *Current Psychiatry Reports, 26*(1), 1–8. https://doi.org/10.1007/s11920-023-01478-w

Wright, K. (2011). *The rise of the therapeutic society: Psychological knowledge and the contradictions of cultural change*. New Academia Publishing/ The Spring.
Wyn, J. (2009). *Youth health and welfare: The cultural politics of education and wellbeing*. Oxford University Press.
Young, R., Kananovich, V., & Johnson, B. G. (2023). Young adults' folk theories of how social media harms its users. *Mass Communication and Society, 26*(1), 23–46. https://doi.org/10.1080/15205436.2021.1970186
Zeng, J., Abidin, C., & Schäfer, M. S. (2021). Research perspectives on TikTok and its legacy apps. *International Journal of Communication, 15*, 3161–3172.
Zinoviev, D., Stefanescu, D., Fireman, G., & Swenson, L. (2016). Semantic networks of interests in online non-suicidal self-injury communities. *Digital Health, 2*, 2055207616642118. https://doi.org/10.1177/2055207616642118

12

Epilogue

Bruce M. Z. Cohen

As editor of the series for which this collection is a part, I tend to stick to the shadows, only making myself visible if authors require some advice, otherwise leaving them to get on with the emotionally and intellectually challenging work of putting such a book together. Natalie and Effie kindly, if maybe a bit unwisely, suggested I could write this epilogue for their volume. There was a vague rationale that, as I have been involved (sometimes as a presenter, more often as an attendee) since the very first Social Production of Mental Health (SPMH) seminar in 2020, I might have something useful to share here. So, taking off my series editor hat, in what follows I offer some brief comments on this volume as it reflects on the current state of Australasian critical mental health research and praxis.

It is highly unlikely that, without the COVID-19 pandemic and the wonders of Zoom, we would have the SPMH seminar series and, indeed,

Bruce M. Z. Cohen (✉)
University of Auckland, Auckland, New Zealand
e-mail: b.cohen@auckland.ac.nz

N. A. Hendry, E. Karageorgos (eds.), *Critical Mental Health in Australia and Aotearoa New Zealand*, The Politics of Mental Health and Illness,
https://doi.org/10.1007/978-3-031-82861-4_12

this book (which is, to my knowledge, the first edited collection on critical mental health research produced in the region). In a short period of time, the seminar series has established itself as a crucial space in Australasia for academics, postgraduates, consumers/survivors/ex-patients (C/S/X), para-academics, and interested practitioners to discuss, debate, and disseminate their critical ideas and research on topics of mental health and illness. Burgeoning out of the darkness of the pandemic, this online forum has become an important meeting point where we can question the taken for granted, speak back to psychiatric power, and work towards solidifying a critical community of activists and scholars in the region. The seminar series has given us critical talks, short provocations, introduced ongoing and newly produced research, launched books, had panel sessions profiling postgraduate research, and has now also given us this volume. Of course, this is not all down to COVID-19 and new online technologies; all that the series has facilitated would have been impossible without the immense organisational and coordination skills of Natalie Ann Hendry and Effie Karageorgos, as well as the original establishment of the series by Baptiste Brossard (who has, since his move to York University in England, set up a similar critical online seminar series for colleagues in European time zones). I would like to thank all three of them for their foresight in setting up and maintaining the SPMH seminar series, which continues simply on the basis of good will and the enthusiasm for critical discussion in a friendly and welcoming environment, rather than through any additional sources of funding (unsurprisingly, there are no mental health foundations, pharmaceutical corporations, or state funders lining up to encourage such collaborations!).

This collection has provided a good overview of the diverse range of critical topics and perspectives that the seminar series has typically incorporated: some have revisited familiar issues with new research (for example, the utilisation of psychiatric discourse as a form of social control in education and the military, as respectively discussed by McLeay and Karageorgos); others have profiled more emerging topics (including social and digital media as spaces for the potential expansion of the mental health industry, as discussed by Hendry and Flore, respectively); and yet some others have proposed new understandings of madness and ways of working with people in extreme states of distress (Sinclair's and Kopua et al.'s chapters, for instance, will come as a surprise to anyone who thinks that being 'critical' in the area means we cannot offer alternative solutions to the practices and 'treatment' options provided by mainstream mental health services). The volume has also demonstrated the continuing importance—as a sociologist, I would say necessity—of utilising social theory to further illuminate patterns of dominance, oppression, and marginalisation in capitalist society which are reinforced through psy-professional interventions. As a key theorist of institutional, medical and psychiatric power, Michel Foucault continues to be a very strong influence throughout the book; likewise, feminist and critical race scholarship remain important reference points for some of the featured analyses; significant is also the emergence of queer theory in a number of the chapters; but perhaps, most of all, it is the influence of Mad Studies scholarship which is proving to have real momentum in our critical space right now, including on a number of critical reflections profiled in the book (for example, in the chapters from The *Re;Minding* Histories Research Group, Sinclair, and Howe). On the latter direction in critical work, I think we ignore (as mainstream mental health services still attempt to) Mad Studies at our peril; guided by the centrality of C/S/X experience and involving a plurality of theoretical directions, Mad Studies is emerging as a natural allay for our critical research endeavours, which includes a growing number of our postgraduate students performing innovative biographical and autobiographical research on madness, labelling, and abuse (emotional, violent, and/or sexual) at the hands of psychiatry.

Perhaps the most symbolic contribution of the book, however, is recognition that thinking critically about mental health and illness is an interdisciplinary pursuit. Challenges to the status quo are as likely to come from psychology, law, and the health sciences as from sociology, anthropology, and history. Despite a management rhetoric in universities that academic disciplines are stuck in 'silos', this has never been the case in our area. In fact, as I am always keen to note in my own presentations, critical research originates as often from those within psychiatry and allied professions as it does from social scientists and fellow scholars outside of the mental health system. Given the relatively small numbers involved in critical mental health research (globally, as well as in

Australasia), it has felt quite natural for us to reach out and collaborate with sympathetic colleagues across the academy as well as beyond. This disciplinary diversity to our critical work can also be found in the background of the authors who have forthcoming books in my book series, who so far come from the areas of anthropology, public health, global studies, social work, counselling, psychotherapy, sociology, and Mad Studies (see for example, Fixsen, 2024; Rose, 2022; Tseris et al., 2024).

In my view, the commonality of the contributors to the present book (and to all who work critically in this area) is their ability to provide understandings, interpretations, and theories which go *beyond* pathology, disease labels, and the current hegemony of the biomedical model (see Cohen, 2025). This is the 'broad church' that I understand as part of the SPMH community and other critical communities globally—these are groups which can problematise, contest, and challenge the common sense reality of 'mental health' and 'mental illness' in society. To a greater or lesser degree, they recognise the following key issues with the mental health industry: first, the historical and cultural fluidity of marking people with the 'mental' prefix (see for example, Cohen & Hartmann, 2021; Metzl, 2009; Showalter, 1985; Summerfield, 2012; Szasz, 1970); second, the lack of validity of the science of academic psychiatry (see for example, Allsopp et al., 2019; Burstow, 2015; Cohen, 2016; Whitaker & Cosgrove, 2015); and third, the lack of efficacy of the practices and treatments offered by the mental health system, including psychopharmaceuticals, electroconvulsive therapy (ECT), and psychotherapy (see for example, Breggin, 1991, 2008; Cohen, 2024; Hahn, 2020; Masson, 1994). Framing research within such a context allows this community to then more seriously address the question of *who benefits* from the establishment, maintenance, and expansion of such powerful institutions which now impinge on our everyday lives.

The result is more powerful—and frankly, more interesting!—research such as that which has been captured in *Critical Mental Health in Australia and Aotearoa New Zealand: Social and Historical Perspectives*. The work in this collection will hopefully inspire others to carry out their research in more dynamic ways, and draw them to our critical SPMH community.

References

Allsopp, K., Read, J., Corcoran, R., & Kinderman, P. (2019). Heterogeneity in psychiatric diagnostic classification. *Psychiatry Research, 279*, 15–22. https://doi.org/10.1016/j.psychres.2019.07.005

Breggin, P. R. (1991). *Toxic psychiatry: Why therapy, empathy and love must replace the drugs, electroshock, and biochemical theories of the "new psychiatry"*. St. Martin's Press.

Breggin, P. R. (2008). *Medication madness: A psychiatrist exposes the dangers of mood-altering medications*. St. Martin's Press.

Burstow, B. (2015). *Psychiatry and the business of madness: An ethical and epistemological accounting*. Palgrave Macmillan.

Cohen, B. M. Z. (2016). *Psychiatric hegemony: A Marxist theory of mental illness*. Palgrave Macmillan.

Cohen, B. M. Z. (2024). *The Assault on Truth* in the academy: Talk therapy and the social control of women. In W. Middleton & M. J. Dorahy (Eds.), *Contemporary perspectives on Freud's seduction theory and psychotherapy: Revisiting Masson's 'The Assault on Truth'* (pp. 165–177). Routledge.

Cohen, B. M. Z. (2025). A critical manifesto for the sociology of mental health: Past, present, future. In E. von Kardorff, M. Harbusch, & D. Robin (Eds.), Zur Gesellschaft der verletzten Seelen: Neue soziologische Perspektiven auf Psychiatrie, Gesellschaft und Subjekt [On the society of injured souls: New sociological perspectives on psychiatry, society and the subject]. Springer VS.

Cohen, B. M. Z., & Hartmann, R. (2021). The 'Feminisation' of psychiatric discourse: A Marxist analysis of women's roles in neoliberal society. *Journal of Sociology, 59*(2), 349–364. https://doi.org/10.1177/14407833211043570

Fixsen, A. (2024). *The construction of eating disorders: Psychiatry, politics and cultural representations of disordered eating*. Palgrave Macmillan.

Hahn, P. D. (2020). *Prescription for sorrow: Antidepressants, suicide and violence*. Samizdat Health.

Masson, J. M. (1994). *Against therapy* (Rev. ed.). Common Courage Press.

Metzl, J. M. (2009). *The protest psychosis: How schizophrenia became a black disease*. Beacon Press.

Rose, D. S. (2022). *Mad knowledges and user-led research*. Palgrave Macmillan.

Showalter, E. (1985). *The female malady: Women, madness, and English culture, 1830–1980*. Penguin.

Summerfield, D. (2012). Afterword: Against 'global mental health'. *Transcultural Psychiatry, 49*(3–4), 519–530. https://doi.org/10.1177/1363461512454701

Szasz, T. S. (1970). *The manufacture of madness: A comparative study of the inquisition and the mental health movement*. Harper and Row.

Tseris, E., Franks, S., & Bright Hart, E. (2024). *Psychiatric oppression in women's lives: Creative resistance and collective dissent*. Palgrave Macmillan.

Whitaker, R., & Cosgrove, L. (2015). *Psychiatry under the influence: Institutional corruption, social injury, and prescriptions for reform*. Palgrave Macmillan.

Index[1]

[1] Note: Page numbers followed by 'n' refer to notes.

N. A. Hendry, E. Karageorgos (eds.), *Critical Mental Health in Australia and Aotearoa New Zealand*, The Politics of Mental Health and Illness,
https://doi.org/10.1007/978-3-031-82861-4

The manufacturer's authorised representative in the EU is Springer Nature Customer Service Centre GmbH, Europaplatz 3, 69115 Heidelberg, Germany. If you have any concerns regarding our products, please contact ProductSafety@springernature.com

Printed and bound by CPI Group (UK) Ltd, Croydon, CR0 4YY
30/09/2026
02225565-0010